NORTH POLE

The Earth series traces the historical significance and cultural history of natural phenomena. Written by experts who are passionate about their subject, titles in the series bring together science, art, literature, mythology, religion and popular culture, exploring and explaining the planet we inhabit in new and exciting ways.

Series editor: Daniel Allen

In the same series

*Air* Peter Adey

*Cave* Ralph Crane and Lisa Fletcher

*Clouds* Richard Hamblyn

*Comets* P. Andrew Karam

*Desert* Roslynn D. Haynes

*Earthquake* Andrew Robinson

*Fire* Stephen J. Pyne

*Flood* John Withington

*Gold* Rebecca Zorach
  and Michael W. Phillips Jr

*Ice* Klaus Dodds

*Islands* Stephen A. Royle

*Lightning* Derek M. Elsom

*Meteorite* Maria Golia

*Moon* Edgar Williams

*Mountain* Veronica della Dora

*North Pole* Michael Bravo

*Rainbows* Daniel MacCannell

*Silver* Lindsay Shen

*South Pole* Elizabeth Leane

*Storm* John Withington

*Swamp* Anthony Wilson

*Tsunami* Richard Hamblyn

*Volcano* James Hamilton

*Water* Veronica Strang

*Waterfall* Brian J. Hudson

# North Pole

Michael Bravo

REAKTION BOOKS

*For my parents, Nicolette and Paul*

Published by Reaktion Books Ltd
Unit 32, Waterside
44–48 Wharf Road
London N1 7UX, UK
www.reaktionbooks.co.uk

First published 2019

Copyright © Michael Bravo 2019

Printed and bound in China by 1010 Printing International Ltd

A catalogue record for this book is available from the British Library

ISBN 978 1 78914 008 8

# CONTENTS

# Preface

Why does the North Pole matter, when most of us will never visit it and know almost nothing about it? In this book I will treat the mysterious power and allure of the North Pole in a way you will not have seen before. I offer the reader a way to understand why the North Pole truly matters to anyone who knows that our home, planet Earth, is a globe. Many of the explorers who dedicated their lives to unlocking the secrets of the poles were not merely the hardy lovers of precarious adventure or the ardent national heroes they have often been made out to be. A lifetime of knocking on the door of the North Pole led explorers themselves to reflect deeply on the nature of their endeavour, and to appreciate that theirs was a personal and moral quest more clearly rooted in paradox and ambiguity. What had been widely regarded as one of the most coveted geographical goals on Earth – perhaps even the ultimate prize – became a metaphor and possiblly also a quest for the nature of geographical knowledge itself.

One of the surprises for me while researching this book was to discover how much time late nineteenth-century explorers like Fridtjof Nansen, Robert Peary and Adolf Erik Nordenskiöld dedicated to studying the work of polar navigators and natural philosophers of past centuries. The very idea of seeking the North Pole so beguiled these explorers that they felt compelled to search for a deeper history of the poles in which their own polar endeavours would make sense. Seldom was there agreement among them about a single definitive or correct approach

The Eden frontispiece to Fridtjof Nansen's *In Northern Mists* (1911).

or method for attributing significance to the North Pole, but they agreed on one thing, and that was the need to place exploration in both a historical and philosophical context. In the nineteenth century, beginning with John Barrow, the architect of the British programme of polar expeditions, researching and writing polar history became integral to their understanding of polar exploration itself.[1]

For that reason, this book does not aim to tell a linear story charting the attempts of successive expeditions to reach the North Pole. Like the threads running through their own lives, these explorers and philosophers recognized that the character of the North Pole, and of polarity itself, was paradoxical. Spatially, when standing at the North Pole, every direction faces south. Temporally, the North Pole is timeless and has to this day no allocated longitude or time zone. This is no coincidence: the North Pole can be thought of as the origin of time because all lines of longitude, which define time zones, pass through the North Pole. Emperors and philosophers through the centuries have recognized the North Pole's special significance as a point that defines global time, but is not itself subject to it. As we shall see in the chapters that follow, this intersection of worldly and mythical time has held a powerful attraction for those concerned with religious or political power.

Answering simple questions about the North Pole proved very difficult in the time of the ancient Greeks, and even more so for our early modern astronomers, mathematicians and philosophers. Was our earthly North Pole a unique point in the universe or just a fictional cartographic point that could be projected on an unlimited number of celestial bodies? How was the behaviour of a compass needle attracted by the magnetic pole linked to the geographical pole? Was the polarity of a magnetized iron needle or a spherical magnetic rock (called a lodestone) operating on the same principle as the Earth's magnetic attraction? Navigators were the ones to pay a very real price for these very abstract but real philosophical questions. The teams of expedition after expedition in the high Arctic struggled to determine their location and bearing when their compass needles began to circle or wander limply and aimlessly, pointing one way one moment, and another way the next. This uncertainty threatened their confidence as reliable navigators and observers, and frequently left them unsure of their bearings in dangerous ice-filled waters. In the chapters that follow, readers will be able to identify with the struggle of navigators and philosophers to make sense of the strange powers of the North Pole.

The story of the North Pole is more like a cosmographic prism than a straightforward story of discovery through the march of time. It has refracted our understanding of the planet on which we live and the quest to master our knowledge of who we are. In this way, we can answer the question that motivates this book: why has the North Pole mattered, and to whom?

The story being told in this book is an original account, much of it based on previously unpublished research. My debts to a number of distinguished early modern scholars in the history of science and historical geography are if anything greater as a result. When I began this book, I originally set out to write about the Earth's geographical North Pole, and also to make some space for the North Magnetic Pole. What I hadn't realized was that there can be no history of our North Pole without first understanding the celestial poles, which for many centuries were viewed as divine and central to the design of the universe. The Earth's poles received scant attention by comparison, and were deemed in Aristotelian thought to be like the Earth itself, a corrupt pale shadow of their divine counterparts. Had history stopped with the Aristotelians, there might have been no book to write. Fortunately, the Neoplatonists of the Renaissance, bitterly opposed to the Aristotelians, viewed the Earth and the heavens as moving harmoniously. The purpose of cosmography in the sixteenth century was to study and understand these harmonies, and it is only then that the Earth's geographical pole began to be an object of special interest. For readers who, like me, anticipated a book that jumps head first into a discussion of early myths about the Earth's poles, I apologize. The story of our terrestrial pole or poles is inseparable from the history of astronomy and needs to begin with the heavens and our earthly harmonies. So important in fact are these harmonies between Earth and the heavens that cosmography is the true legacy that has survived from the Renaissance into the present day, albeit greatly transformed and reappearing in strange ways. For that reason, the route taken to the North Pole through the chapters of this book will be unexpected, with twists and turns, for readers not well versed in early modern history.

The mysterious power and allure of the North Pole is the common theme running throughout the chapters of this book. It has long been a powerful symbol as far back as the civilizations of ancient Greece, Egypt, India and Persia. We might expect that once the eminent geographers and astronomers of centuries past had defined and mapped the North Pole, the mystery would give way to certainty and even to banality, but history shows that quite the reverse happened. The more attention the North Pole received, and the more men and women of science tried to pin it down, the more its nature proved to be puzzling and beguiling. Hence this is the story not just of the single point on the globe at 90°N, which at first glance seems fixed and unambiguous, but of other North Poles, some more and others less important, that together have been a perpetual source of mystery and paradox, evading all efforts to finally tame them.

The first step is to understand that, short of teleporting to the North Pole, philosophers, scientists, armchair travellers, explorers and writers of popular fiction have long believed that the only way to reach it is by journeying to it. Historically the Earth's geographical pole has been approached in four ways: from land, from the heavens or sky, from under the sea or from the Earth's interior. Some of these journeys have made use of the genius of science or intrepid technological experiment. Others have been fantastical and even utopian stories of incredible or monstrous adventure. In all their variety, what the theme of *journeying* or *approaching* tells us is that the North Pole has never been an entirely isolated point; it is only meaningful in relation to the globe as a geographical system. This is a very good thing for this book because so very few of the historical expeditions attempting to reach the North Pole attained their goal.

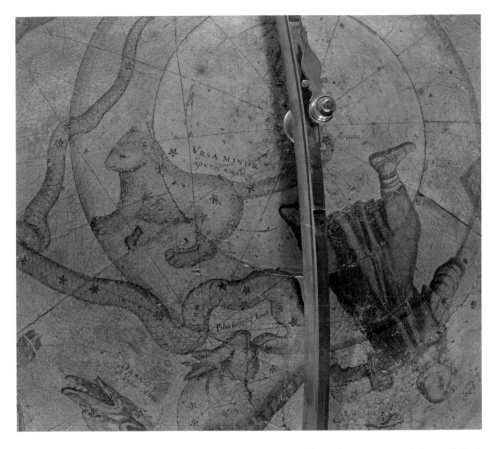

On this beautiful celestial globe made by Emery Molyneux (1592), the star on the end of the bear's tail in the constellation Ursa Minor is labelled 'Alrucaba', more commonly called Polaris, the celestial Pole Star.

Until the twentieth century, each and every one of them failed. Fortunately for the protagonists and their backers, the stories of polar expeditions were really stories about approaching the pole, and in this they fared much better.

Long before people set foot anywhere near the North Pole, artisans and experimenters worked out that they could learn a great deal about the North Pole by literally making models of it. This included maps and globes, which were for the artisans of the Renaissance a very important kind of mathematical or philosophical instrument, as we shall hear more about in Chapter Two. Thus great thinkers, by making charts and other curious instruments in their workshops, were for the first time able to hold the North Pole in their hands and to gaze upon it and the rest of the globe as if looking down on it from far above, emulating a kind of divine insight. Much of what is discussed in the following chapters is a series of accounts of the ingenious, clever, skilful and often mistaken but insightful ways in which the pole has been gazed upon.

Looking down over the North Pole meant much more than just locating it. To do so was to gaze at the whole Earth, to see it in its unity, as an integral whole, and to know it as part of a much larger universe. For ancient Greek astronomers from Aristotle onwards, the Earth lay at the very centre of the universe, entirely still, while all the planets moved in a perfectly ordered procession around it. This giant axis stretched from the celestial Pole Star, high in the heavens, fixed almost directly above the Earth, straight down through the North and South Poles, and the centre of the Earth. Ancient astrologers would seek to discover harmonies between the divine movement of the heavens and the Earth. Any astrologer worth his salt was on the lookout for divine conjunctions of constellations and stars, and omens or portents of dangers ahead. Hence the importance of the geographical North Pole came about first because of the celestial North Pole and its pole star, and our knowledge of the Earth's grid of latitude and longitude was a projection derived from mapping the celestial realm. Without the celestial pole, it made no sense to speak of a geographical pole. Together these poles defined the alignment of the universe, and thus were distinct but inseparable.[2]

For Greek and Arab astronomers, poles were at the heart of the architecture of the entire cosmos. Without poles there could be no geography and crucially no system of orientation for navigation. Living in the twenty-first century, thanks to powerful technologies of aviation and remote sensing, we are accustomed to seeing images of the Earth from high in the atmosphere and outer space.[3] For philosophers of the enlightenment like Kant, the human condition was one of being anchored to the Earth, like ants unable to escape the limitations of a field of vision placed very close to its surface. In Chapter Three we shall discover that the North Pole played a key role in some of the most important innovations in developing instruments of navigation to gain mastery of movement over the surface of the globe. Thus the North Pole provided one of the main keys to help unlock the basic question of human orientation – to know where we are at any moment in time and to know on what course we are heading.

The Celestial North Pole is the point around which all the other stars appear to rotate. With precession, the position of the stars in the night sky actually shifts over time and the nearest star to this centre, the Pole Star, also changes position. In this recent image, Polaris is the Pole Star.

Despite the enormous efforts made by our predecessors to map the globe very precisely, the actual experience of navigators on the high seas has often been the exact opposite – disorientation. Navigation has historically been an art of developing techniques and instruments to find one's way out of a blind corner when confronted by uncertainty, or by finding oneself lost. Only in the last forty years have global positioning satellites (GPS) enabled us to feel we can always know exactly where we are when journeying across the tundra or sea (provided one has a good clear view of the sky and a satellite receiver with charged batteries!).[4] The secret of navigation through the ages

has been to know one's place in relation to another known place or places. Disorientation happens when this relationship fails us. The compass needle helps us to know where we are in relation to the magnetic North Pole through the power of attraction of its opposite pole, but the geographical North Pole militates against this idea of relative position. Isn't the North Pole a given fixed place in the universe, like the Pole Star in the firmament high above, and absolutely correct in its own right? Should it not therefore be independent of other places and immune to illusions that create disorientation? This is a question we will return to.

Historians of exploration know all too well that the pages of navigators' journals, particularly of those in high latitudes, attest to fraught periods of distress and disorientation. This turns out to be no accident or coincidence. The most obvious and ancient example of a polar-oriented instrument, the compass, consistently failed to give reliable bearings at high polar latitudes. Establishing places and locations in a region where the lines of longitude crowd in on each other bothered globe makers like Gerardus Mercator (1512–1594) as much as Royal Society-entrusted navigators like Thomas James (1593–1635). A practical method for calculating a ship's longitude at sea to within a tolerable accuracy was not devised until the second half of the eighteenth century.[5] Even afterwards the nature of the geographical pole and the source of its companion pole's magnetism remained mysteries, and it was completely unclear whether magnetism was a force or a fluid flowing from two or four poles, from the Earth's interior, or from somewhere else.

The paradox explored in Chapter Three is that just as philosophers or scientists discovered how to map the poles, they discovered that poles were not simply points on the Earth's surface after all. Magnetism and the discovery of polarity turned thinking about poles on its head. Where there had been one pole, the Elizabethan experimentalist William Gilbert (1544–1603) showed there were multiple poles. Where poles had safely been assumed to be located on the Earth's surface, they now seemed to emanate from the Earth's interior or other planetary bodies. Just as new instrument designs could point north with

Time-lapse photograph of stars seeming to turn round the Pole Star, 2010.

increasing accuracy, it was increasingly unclear what they were pointing towards!

Arctic polar explorers who in their youth had been intent on reaching and thereby conquering the North Pole – Nansen (1861–1930), Peary (1856–1920), Nordenskiöld (1832–1901) – in later years became polar historians. In Chapter Five we will see how they became historians of an enterprise that was personal and autobiographical, and which simultaneously purported to be the story of Western science and civilization. These histories sought to lift the veil on an Arctic shrouded in the mists of time harking back to a lost Eden. Mystics and utopians looked beyond the borders of Europe in seeking a universal story about the origins of the human race. Ancient Persian and Hindu texts inspired theosophists and mystics who claimed to have hidden knowledge of a lost Aryan homeland at or near the North Pole, whose people took flight with the arrival of a polar ice sheet and settled the northern lands of Asia and Europe. Utopias and grand theories about the Earth's hidden forces were also fair game for satirists who delighted in puncturing polar mysteries by mocking the interests, pretensions and reputations of anyone who aspired to rule the North Pole. Whether satire can reasonably be thought of as a source of everyday philosophy is debatable, but what becomes clear in this chapter is that satire was a valuable, even indispensable, means of making public difficult questions about what was at stake in being able to speak authoritatively about the North Pole.

Those readers already familiar with polar history will notice how many important polar explorers, men and women, of all races, are scarcely touched on in this book beyond a mention here: for example, the remarkable achievements of explorers like George Nares (1831–1915), Karl Weyprecht (1838–1881), Salomon Andrée (1854–1897), Walter Wellman (1858–1934), Matthew Henson (1866–1955), Ivan Papanin (1894–1986), Wally Herbert (1934–2007), Will Steger (*b.* 1944), Liv Arnesen (*b.* 1953), Ann Bancroft (*b.* 1955) and Pen Hadow (*b.* 1962). This is a short book, and they scarcely get a look-in, but they are important. They are just a few of the people for whom the North Pole has been a source of reflection on the human condition of inhabiting the globe. The diversity of their thinking is as important as the significance they attribute to polar journeys. Although their feats are invisible in this short book, I like to think they are present as interlocutors in the philosophical conversations that over the centuries form the real heart of the story of the North Pole.

of the entire globe. For emperors, the fixed star was an intensely important symbol of their worldly power and its heavenly authority.

For the Inuit people, the *sila*, a word that encompasses sky, heavens and air, is very central to their lived experience. The stars and constellations criss-crossing the sky are part of the *sila*. Their trajectories across the night sky mark out trails or pathways, not unlike the precise network of connected trails that define the map of Inuit routes across the sea ice. In this world, trails not only connect places where people live; to be 'on the trail' is itself a way of being at home.[1] Thus movement in Inuit culture is all-important in understanding all aspects of being throughout the cosmos. Within that world, certain stars, individually or grouped in constellations, provide Inuit travellers with skymarks (like landmarks) and trail markers to help them orient themselves while on their trails on the tundra or across the sea ice.

'Connectedness', a quality that permeates the fabric of Inuit society, is at once geographical and emotional. Just as it binds people and places together, so too the night sky forms part of this intimate matrix of connection. To be safe and healthy requires possessing precise knowledge about one's environment. The vocabulary of place requires learning hundreds of place names on the land, sea, ice – and in the night sky. As hunters or families travel trails across the sea ice or land, they reorient themselves using a succession of shifting horizons.[2] The shape and texture of each horizon is unique to a moment in time and space depending on the weather conditions, time of day or season, as well as the view from a given location. Decoding the shape of shifting horizons depends on learning stable unchanging landmarks and their corresponding place names. As we shall soon see, the Pole Star is one such landmark, except that, placed high in the heavenly sphere, the term 'skymark' is more apt.

The knowledge required to navigate a system of trails is encoded in narrative stories. Typically, a young person will have listened carefully to an experienced traveller, and then remembered their description of a journey and details of a trail followed.[3] Sometimes, the story of a trail will involve or take

Pairngut Peterloosie, a distinguished Inuit elder of Pond Inlet, in 2011 documenting and reminiscing about key trails from the time that her family lived in camps on the land. Inuit women like Pairngut travelled thousands of kilometres and had a very extensive knowledge of their territory.

the form of a myth or legend that may give meaning and shape to a trail. Similarly, the trail of a constellation that can be tracked across the night sky is described in a myth that explains this movement. However, rather than thinking of these myths as a stationary form of mapping, one wants to keep in mind that Inuit travellers are using a moving frame of reference that changes as they themselves move. If one thinks of the Inuit world in terms of the fluid movement of people, animals and spirits across intersecting or connected systems of trails, one begins to gain a better sense of how the stars figure in their navigation tradition.[4]

The ever-changing network of star trails covering a vast night sky is tamed or made manageable by knowledge of the precise movement of constellations. This is no less true of the Inuit landscape, covering a huge swath of land, sea and ice, but encoded as very precise, tightly knitted knowledge of places, trails and people. The *Pan-Inuit Trails Atlas* of the Inuit world of Arctic North America conveys the sense of a space that is both vast and intimate, linking places as far away as Alaska and Greenland. For people of the sea ice, trails, horizons and place names are written into the contours of the world.

The scale and reach of the Inuit world means that the visible night sky looks different depending on one's latitude. The Inuit people dwell on the coasts of three oceans (Atlantic, Arctic and Pacific), spanning almost 25° of latitude – from 55° (Labrador Coast) up to about 80° (Northwest Greenland and the Arctic archipelago of Canada). Across this expanse, there are very pronounced contrasts in the topography of the land, the presence of sea ice, and the dialect and vocabulary of people. Yet

for all of these differences, their world is held together by stories containing a complex knowledge of place: intimate, detailed and connecting.[5]

How many centuries back one can reliably project Inuit oral knowledge traditions is very hard to pin down. However, the historical migration or movement of people, studied by archaeologists, documents successive waves of migration contemporaneous with developments of navigation in other civilizations. The high Arctic of Canada and Alaska was populated through a series of migrations from Asia on the one side and from Greenland on the other. The migration of the 'Independence II' palaeo-Eskimoan people into Northwest Greenland about 700 BCE was roughly contemporaneous with the earliest Presocratic

The Pan-Inuit Trails Atlas shows the intricately connected network of trails spanning the North American continent from Alaska to Greenland.

Greek philosophers in Miletus in Anatolia.[6] This was preceded by a still earlier 'Independence 1' migration (*c.* 2400 BCE) some centuries before the development of the Minoan civilization in Crete (*c.* 2000 BCE). Thus the peopling of northern Greenland by hunter-gatherers, like the early Minoan settlements of Crete, was the work of skilled maritime cultures. It would be nearly another two millennia before the Greek astronomers Hipparchus and Ptolemy put in place the foundations of a world view based on the globe with latitude and longitude.

The most detailed and reliable study of Inuit star lore comes from Igloolik (68°N) in Northern Canada, where a society of elders worked closely with anthropologist John MacDonald, author of *The Arctic Sky* (1998), who recorded, codified and reflected deeply on their knowledge of the chief stars in the celestial heavens.[7] The Pole Star (part of Ursa Minor in Greek astronomy) has a place above Ursa Major (Tukturjuit, the caribou). For the Inuit, some of the most important stars are Aquila (Aagjuuk, two sunbeams marking the end of the dark period of winter solstice), and Arcturus and Vega (Kingulliq), each widely used for some form of orientation.

The Pole Star is significant and exceptional as a fixed star, useful for orientation, even as the concept of 'north' is foreign to

Joanasie Makpah and Pairngut Peterloosie sharing memories and life histories along carefully delineated trails crossing Baffin Island, 2011.

traditional Inuit knowledge. Inuit names and meanings vary with dialect. *Turaagaq* from Northern Quebec means 'something to aim at', referring to its orientation value.[8] Also widely used, *Nuutuittuq* from Igloolik, means 'never moving' or fixed, suggesting stillness and constancy in the sky. To what extent Igloolik Inuit are guided by this star involves some geographical subtlety.

Nuutuittuq is particularly useful when setting out after a night at camp. Changes in weather, like a blizzard, may completely alter the appearance of a landscape overnight. A fall of fresh snow could cover the trails or a change of wind direction could alter the contour of the guiding snowdrifts. Thus one hunter, Panikpakuttuk, explains that when travelling with archaeologist Graham Rowley, they would plant a stick in the ground pointing to Nuuttuittuq before going to bed, and use it in the morning to regain their bearings before setting off. By starting in the right direction, a hunter could then follow the trail by the more usual methods of orientation using the wind, snowdrifts, landmarks and existing trails.[9]

Emil Immaruituq, an Igloolik Elder, training his dog team, 1988.

Knowledge was often shared and passed down through the generations, but orientation could also be learned through patient observation and curiosity. One elder, Abraham Ulaajuruluk, recalled a simple experiment he had once carried out to satisfy his curiosity piqued by the one star that seemed to remain fixed in the night sky. One evening before going to bed, he positioned 'a harpoon shaft pointing directly to this star to see if it would in fact move' when morning had come. He 'discovered that while Tukturjuit (meaning Caribou, or Ursa Major) had changed its position completely . . . the harpoon was still pointing at this [same] star'. Reflecting on this, he concluded that 'I had discovered the stationary star, *Nuutuittuq*!'[10]

In truly dangerous situations that present a real risk of suddenly becoming lost, the Pole Star at night time could save lives. For example, Herbert Amarualik pointed to the perils of 'navigating on moving sea ice . . . [when] it is breaking up and being driven by currents'. People have died in situations like this, and expertise, judgement and patience were all recognized as necessary for survival. At such a moment, if the prevailing wind, normally the most important method of taking a bearing,

Well-trained dogs acquired a remarkable ability to read the sea ice and follow a bearing for hours over considerable distances. Igloolik, 1988.

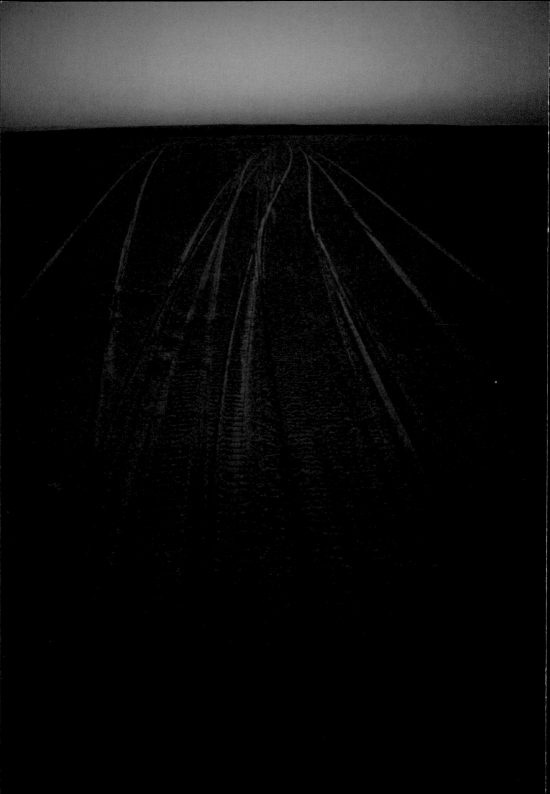

died to a calm, with a clear sky, one could trust the fixed star to work out the direction of land and return to safety.[11] By revealing the lay of the land on a large scale, the Pole Star could be the difference between life and death.

## Trusting in the movement of the night sky

Once a sledge party had set out in the right general direction for their destination and needed only to stay on course, the fixed position of the Pole Star proved less helpful. While driving a dog team, a visible star was only useful if it was within the driver's field of view. Having to raise one's head, or look off to the right, left or behind, meant losing concentration and taking one's sight off the conditions of the trail. This fetched too high a price for such stars to receive much use.[12] The Pole Star appears high in the sky in the Inuit world, well above the horizon, and thus too high to be practical as a bearing. Igloolik, at a latitude of 70°N, lies at the outer limit. At lower latitudes, following the star could be manageable; further north than 70°, the star is so high that even trying to take one's bearings from it is judged difficult and impractical. In the very northwest of Greenland, for example, where Robert Peary established his base for his attempt on the pole, the Pole Star was unnamed, seemingly anonymous.[13] Even for George Kappaniaq of Igloolik, who identified Nuutuittuq as being at the centre of *sila*, the sky, this high elevation gave the Pole Star only limited importance. Thus fixity and centrality, qualities that will be revealed in the next chapter to be of supreme importance in Western astronomy, were only of limited practical importance for Inuit. For this reason, there was no legend or myth surrounding Nuutuittuq.

Even in those places where Nuutuittuq was used, the Sun, Moon and other stars were at least as important. It is clear from the work in Igloolik that Inuit traditional knowledge of some stars and constellations was rich, recognizing in them seasonal movements across the sky, rising and setting. The movements of some of the stars, the Sun and the Moon were learned through the stories that explained the spiritual relationships and shared

A seal-hunting journey in mid-afternoon on a November day. The subdued light nevertheless illuminates the detail and texture of the sea ice trails.

histories binding their journeys across the sky. This helped people learn to remember where to find the stars and to track their paths through the night sky.

The story of the relationship between Arcturus and Vega is instructive. The version told at Igloolik reveals how the interplay between murder and vengeance reduces a vast landscape or night sky to a tightly knitted trail or pathway. A nasty old man named Uttuqalualuk, who has murdered his brother-in-law, endeavours to keep it a secret from those around him. Taking pleasure in others' grief, when he sees Iliarjugaarjuk, an orphaned boy living with his grandmother, he taunts the child, telling him to eat the meat of his dead mother's tail bone. The boy tells his grand-mother, who knows the old man's secret, and urges the boy to use his knowledge of the murder to confront the old man. Reluctantly

Knowledge of Inuit astronomy is indebted to a multi-year traditional knowledge project led by Inuit elders, hosted by the community of Igloolik's Research Centre, shown here.

and afraid, the boy does so, but for his pains is chased by the old man around his igloo. The boy's grandmother, Kingulliq (or Kingullialuk), gives chase to rescue him, but arrives too late at the scene to catch the old man, who has fled with the boy. The three of them are said to rise up into the sky: the old man and the boy becoming the double-star Arcturus (comprising Arcturus and Murphid), and the grandmother chasing behind them is Vega. In this story there is no refuge from vengeance, as the feud is forever enacted across the night sky by two of the brightest stars, Arcturus and Vega.

The relational character of the story of Arcturus and Vega nicely illustrates a key feature of Inuit cosmology. The morality of violence and revenge that keeps these stars locked in perpetual motion enables their track and position across the night sky to be used as a clock, to give an indication of time during darkness. The story that unfolds from the faithful boy and his grandmother confronting the old man's treachery becomes one of the most trusted sources of astronomical orientation.

## The star of empires stands aloof

Nuutuittuq, the Pole Star, for all its stillness, holds no special metaphysical significance for the Inuit. This is in sharp contrast to the cosmographical traditions of the ancient Greeks, medieval Arabs and Renaissance astronomers, who celebrated the celestial pole for its position on the central axis of the universe, with all heavenly bodies rotating around it. Explorers at the turn of the twentieth century, dubbed the 'Heroic Age' of polar exploration, were well versed in cosmography. Even if they knew that the Earth orbits the Sun, rather than vice versa, the symbolic power of the Earth's poles retained its cosmographical status as a place unique in its purity and deserving of veneration. In that sense, the explorers of the Heroic Age were followers of the ancient cosmographers who understood the North Pole in a mystical or ethical sense, and believed that to approach the geographical North Pole was to come closer to an immanent experience of the divine heavens.

The single name most widely associated with the North Pole is Robert Edwin Peary, who on 6 April 1909 established Camp Jesup, which he claimed was within 8 kilometres (5 mi.) of the North Pole, thereby claiming its discovery. On Peary's return to the United States, his former ship's surgeon turned rival, Frederick Cook, announced that he had in fact beaten Peary to the pole, reaching it in 1908. A bitter controversy ensued that lasted for many years as the two men struggled to remove the stain of accusations from their character and to prevail over the other. In such circumstances, characters can begin to look one-dimensional and it becomes easy to overlook their profound if flawed inner lives.

Historian Michael Robinson encourages us to remember that polar explorers in Peary's time deliberately cultivated a highly nationalist, masculine and heroic image for the press and the public. So overwhelmingly important was publicity for fundraising on the lecture circuit that the patronage of newspapers and the press played a central role in organizing the expeditions in the first place.[14] Peary's construction of himself as a rugged, self-sufficient man exuded a manhood that held a strong appeal to readers in a society that was anxious about rapid urbanization and possessed a strong sense of loss and nostalgia for a more heightened masculinity, allied to a sense of the primitive unfettered by the excessive refinements of society.[15] As geographer Karen Morin observes, this masculinity appealed to the world of Peary's backers, particularly the wealthy, commercially minded and avuncular Charles P. Daly, President of the American Geographical Society (1864–99), who did so much to promote Peary.[16] Knowing that polar exploration in this period epitomized this civic need for a heightened sense of masculinity can help us to understand that Peary's beliefs about the North Pole, though doubtlessly heartfelt, were to some degree scripted for an audience hungry for tales of heroic purity and suffering.

The idea to which Peary was beholden, that the celestial Pole Star and its terrestrial correlate possessed a quality of absolute purity and divine authority, would have been philosophically absurd to Inuit ways of thinking.[17] Arcturus and Vega were well

Drawing of the
constellation
of Herakles (or
Hercules) in the
*Atlas Coelestis* (1729)
by John Flamsteed.

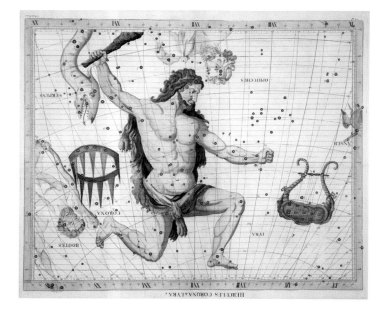

known to most trained navigators because of their brightness in the night sky. Peary had studied for a degree in civil engineering at Bowdoin College, Maine, and went on to develop expert navigation skills. The constellations of the Arctic night sky (in the winter months when they were visible) were as open to him as a book of classical mythology. He fashioned in his own image a constellation between Arcturus and the pursuing Vega – the gigantic Herakles (upside down) whose feet are planted high in the night sky, forever encircling the Pole Star. In identifying himself with the image of Herakles, he was only doing what imperial figures before him had done time and again in a tradition stretching back to the ancient Greeks.[18]

Any imperial expression of a desire for universal earthly authority requires an act of sovereignty that breaks a taboo in order to become the law-maker rather than a law-follower. For Peary, to reach the pole was to breach it, to conquer it and to claim the source of time itself, the revolution of the Earth. In seeking this absolute end, Peary confessed to having sacrificed some essential part of himself in pursuit of this ambition. Reflecting on this for the readers in prose carefully crafted in his published narrative, he wrote that the quest had 'become so much a part of my being that, strange as it may seem, I long ago ceased to think of myself save as an instrument for the attainment of that end'.[19] Embracing the mythology so completely, he had come to inhabit it. The idea of the North Pole was strangely home to him.

Imperial claims to universal authority have always required the assent of the conquered or colonized through art, rhetoric and mythology. For that reason, the stories that Peary tells about the universal authority gained from claiming the pole are in fact always relational. In ruminating on the theme of Antaeus, after reaching the pole, Peary acknowledged that 'the grim guardians of Earth's remotest spot will accept no man as guest until he has been tried and tested by the severest ordeal.'[20] Coming through this Heraklian ordeal is the source of legitimacy for planting the Stars and Stripes in the middle of five flags at the pole, a display of patriotism and patronage. The ritual of taking 'possession of' nothing less than 'the entire region, and adjacent' is witnessed, in Peary's words, by 'my dusky companions'.[21]

Peary took navigational competence as a measure of the hierarchy of the imperial order his expedition represented. The celestial knowledge of the Inuit stacked up unfavourably alongside 'the Arab shepherds' who recognized the Pole Star. His comparison with Arab shepherds was no coincidence. Similarities between Arctic snowscapes and Bedouin desert landscapes were striking. That the landscapes of snow/ice and sand – the snowdrifts and sand dunes – were shaped by prevailing winds was recognized by the most experienced travellers. Similar techniques (up to a point) based on the winds could be used for reading snow and desert landscapes and for setting a course and following a bearing. Thus the comparison between Inuit hunters and Bedouin shepherds possessed a certain logic.

Being at such a high latitude for an attempt on the pole made Inuit star knowledge seem, ironically, less competent in Peary's eyes than it had when further south. Unlike the Mediterranean pastoralists, observed Peary, the Inuit 'have not noticed that one star is the centre about which all the others move, nor have they

Drawing of Ursa Minor, the 'lesser bear', in Al-Sufi's *Book of the Fixed Stars* (c. 1010). The table lists the names and positions of the stars.

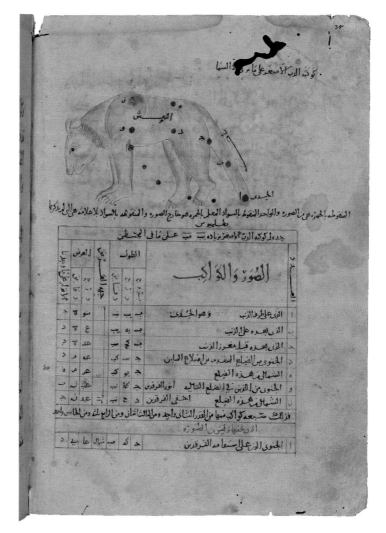

set apart the planets, which to them are simply large stars'.[22] Failing to recognize the Pole Star, in essence, implied that the Inuit, no matter how astute in reading time from the stars, were rulers of their locality, but by no means masters of the globe.

Peary's comparison, though loaded with what are ultimately unfounded imperial assumptions, is important to unlocking his thinking or charter for the significance of the pole. His standard for indexing 'cultures of the pole' rests on a distinction between *wayfinding knowledge* and *scientific navigation*. Wayfinding is an experiential means of finding one's way through using one's body and senses to build a relationship with one's environment, whereas scientific wayfinding follows a more rule-based set of methods using observation, rational calculation and measurement.[23] Both are critical sets of skills in exploration, so much so that Peary served his apprenticeship in Arctic wayfinding with Inuit experts: learning to read the hidden meanings in the shape of the snowdrifts, the shifting winds, the textures and colours of the sea ice and ice ridges, and much besides – very similar to the skills acquired by shepherds in reading the hidden meanings on the surface of desert landscapes, or that sailors acquire through the apprenticeship and constant practice of observing the sea and sky. Across all of these cultures, the repertoire of wayfinding skills includes the naming, recognition and use of the stars, constellations and planets. Were Peary to have used Inuit guides from the lower latitudes, he might have rated them more highly.

The universal authority Peary invests in his achievement in scientific navigation is something he inherits, or more accurately borrows: a distinctly early modern tradition of practical mathematics, instrument-making and astronomy, which was in turn indebted to the astronomers of the Assyrians, Greek, Hindu and Arab civilizations. In that sense, Peary's imperial vision of the North Pole is based on Greek and Arab systems of global geography based on astronomy. His condescension towards Inuit star lore was only slightly tempered by his considerable dependence on their environmental knowledge, navigational experience and the hospitality shown towards him and his American travelling companions. One needs to remember that Arab astronomers played a critical role in the development of medieval astronomy, knowledge that eventually served as a foundation for Renaissance astronomers in Europe. Those same eminent Arab astronomers like Al-Sufi (903–986 CE) were indebted to Bedouin peoples for many of the star names and lore that became the stock-in-trade of Arab and later European astronomy. Thus traditional wayfinding knowledge repeatedly informed the development of astronomy, while being romanticized or disregarded as a kind of poor cousin, an inconsequential local knowledge.

Thus when Peary gazed polewards in the winter night sky off the north shores of Greenland, he fully appreciated that his desire to stand directly beneath the Pole Star, to envisage his future presence there, albeit in daylight, was an act of philosophical imagination. The Pole Star is only fixed in appearance. In reality, the night sky was shifting ever so slightly, following a cycle lasting about 26,000 years – what astronomers since Hipparchos have termed 'precession'.[24] In practice, this means that at some moments in history, the night sky directly above the geographical North Pole has been occupied by different stars, and at some moments, by no star at all. For example, on the night of Jesus's birth, Polaris was almost directly overhead at Etah at 78°N (Peary's Inuit base); at the death of the great Alexandrian astronomer Ptolemy, the same star was at 79°N; at Muhammad's birth, it was slightly closer at 81°N. By 1900 the Pole Star was almost at the zenith, directly over the geographical pole.

In the past and future, different stars will have had the honour of being the star nearest the pole. At the time the Pyramids were being built, the Pole Star was Thuban; in 8,000 years' time it will be Deneb, and eventually Vega. Only in recent centuries has Polaris been the star most directly overhead, less than one degree away from the polar axis. In truth, no star remains in a fixed position over a time scale of millennia; instead, it is the deep-seated need in human beings to seek fixed reference points that seems to remain constant, rather than the reference point itself.[25]

Determining who had reached the sacred pole would ultimately depend on a combination of expert tribunals and the news-reading public in metropolitan centres. They would adjudicate on the critical question: by what standards would the observations of Robert Peary or his African American co-traveller, Matthew Henson, be judged to count as having been there? Peary's philosophical goal – to stand on 'the North Polar axis of the earth'[26] – was almost impossible to verify except within a margin of error of about 16 kilometres (10 mi.), because of the difficulty of using instruments on the moving sea ice to measure the altitude of the Sun. Henson's entitlement to claim similar priority of discovery was couched in terms of hierarchies of the expedition's leadership as well as race. Today Matthew Henson's achievements, as well as those of the Inuit of Etah, are justly celebrated alongside those of Peary.[27]

## The pole star among the Greeks

Understanding the spatial boundaries of the North Pole, where it begins and ends, is not as obvious as one might at first suppose. Of course the measurement 90°N pins it down with total precision. However, ambiguity creeps into the question in two ways. In common usage over the last two centuries the North Pole has been adopted and used – one might say misused – to add colour to travel accounts of voyages that were merely heading northward into high latitudes. The goalposts have been shifted over time by sensational exhibitions, publishers, instrument makers, geographical societies and explorers themselves. The slipperiness of the term reflects different interests at play in the construction, disputation and celebration of polar exploration. In political terms, the idea of the North Pole resonated with national aspirations for empire-building and colonial expansion into the Arctic. When Peary arrived at the place he deemed to be the North Pole, he believed he was entitled to claim the entire region around the pole for the United States, even though his president, William Taft, decided otherwise. Thus the shifting boundaries of the North Pole as a region or frontier were profoundly historical in their definition, reflecting the changing values and ideas of nation-building and empire. Disputes over success and priority – who reached the North Pole first, and whether it was truthful and close enough to count – turned on standards of evidence and the precision of scientific instruments, both of which were also bound up in the politics of nationalism and international rivalry. To make sense of the claims made about the North Pole in the name of science and nationalism, we have to keep one eye on the pole itself and the other eye on the inhabited temperate zones from where navigators and their patrons planned, executed and ultimately made the judgements about geographical discovery. Readers will begin to see in the coming chapters that the North Pole has occupied a very special place in the geographical imagination of the globe.

To gain a better perspective on the history of the pole, it helps if we go back in time more than two millennia to Greece in late antiquity, when there was no expectation of human beings being able to sail to the North Pole. A fascination developed with tracing the expansion of Europe's northern frontier to the reputedly furthest inhabited settlement in the world, a semi-mythical place called Thule. A frontier is by definition a geographical boundary or edge. Archaic Greek oral traditions had once famously marked the boundary between the inhabited Mediterranean and the unfathomable Ocean beyond by constructing the Pillars of Herakles at the western entrance to the Mediterranean, one of the wonders of the Ancient World. The story of Pytheas' voyage to Thule in 325 BCE is one of the voyages that moved the boundary of the known world further out and northward. The location settled on by Ptolemy was about 63°N, and by the late nineteenth century this was variously identified with Iceland, Norway and the Shetland Islands. Its resonance was that it represented the imagined limit of Europe's northern empires and colonies.[28] This famous expedition actually sailed in the fourth century BCE from what was then called Massilia (Marseilles), a thriving Mediterranean port in a well-established trading network between Rome and its allies. Commodities sent south from the interior of Gaul passed through Massilia, where they were loaded onto ships destined for other Mediterranean coastal towns and their entrepôts.

Pytheas, astronomer and navigator, would be lost to us were it not for fragments of records of his periplus (sailing directions) that survived and were studied by Hipparchus. In Pytheas' time, navigators didn't have the benefit of any star sufficiently close to the zenith to count as a Pole Star. The solution he is thought to have come up with was practical and savvy. He made use of a group of three stars relatively close to the celestial pole, describing them, intriguingly, as forming the shape of a quadrangle, not a triangle as one might expect. The location of the celestial pole was pictured or imagined as a missing fourth star, making a corner of the quadrangle, relying on the geometric shape to project the missing star. Measuring where the Pole Star ought to be by taking the elevation (angle) of one of the three stars in the quadrangle gave Pytheas a direct approximation of his latitude.[29]

It seems paradoxical that cultures most renowned for undertaking long-distance navigation, requiring great practical knowledge and skill, have used imagined positions and trajectories of stars. In the Micronesian *etak* system, for example, navigators imagine themselves to be stationary while stars and constellations follow trajectories rising and setting over the horizon. This may not be so strange in that the ability to navigate by incorporating the heavens and horizon into a shifting spatial framework reflects highly sophisticated abstraction. The Pole Star viewed in tropical latitudes of Micronesia or Polynesia is low near the horizon and can serve either as a bearing or an indicator of position when it disappears beneath the horizon.[30]

Such spatial systems, however ephemeral or basic they may seem to the novice, cannot be made out of thin air; maritime societies that have spawned traditions of long-distance navigation

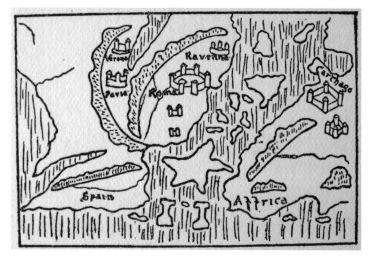

The Pillars of Hercules shown on a 10th-century Anglo-Saxon map.

out of sight of coastlines (including the Phoenicians, Greeks, Melanesians and the Inuit) have patiently accumulated the experience of observing the patterns of movement in the heavens, year on year, for centuries. These cultures have also possessed great traditions of oral history with which to encode astronomical observation and corresponding rules and techniques of navigation, making them portable and equipping their mariners with the means to navigate.

For the ancient Greeks, a seafaring people, the celestial realm was found in the harmonious movement of the stars across the sky celestial, a system that functioned as a reliable source of truth of the highest importance. Ptolemy described astronomy as a 'self-sufficient thing', meaning that its authority wasn't contingent on human or other systems. The movement of the heavens functioned as an independent source of truth. In his view this autonomy made it the 'loftiest and loveliest of intellectual pursuits'. Studying the heavens brought him closer to the divine; in other words, for Ptolemy astronomy was a personal and ethical undertaking. Strange as though that may sound, it is a critical insight because it enables us to understand why for Ptolemy the two North Poles (celestial and ecliptic) of the heavens are a completely reliable source of true knowledge, against which other stars can be measured and positioned.

The celestial pole in Ptolemy's world was positioned high on the central axis of the Earth and universe, where the axis intersects the outermost shell of the planets. It is this pole that is directly above the Earth's geographical poles and that matters most for this story. The Earth's equator or plane of rotation, however, is at an angle to the plane in which the Earth orbits the Sun. Another way of thinking about this is to imagine that in Ptolemy's system, where the Earth is taken to be stationary and not rotating at all, the ecliptic is the perceived path of the Sun across the sky. The plane of the ecliptic has its own pole (the ecliptic pole). It made sense to Ptolemy to identify the position of stars by measuring their angle from the ecliptic pole. Charts and globes of the night sky have therefore traditionally been made with the ecliptic pole as the central reference point.

Ptolemy's ethics have been traced back through multiple strands from earlier philosophers in antiquity, particularly Plato five centuries earlier. In the *Timaeus*, Plato had argued that heavenly bodies are gods, and that they in turn possess a generative force and create all other living things in the universe.[31] Advancing a slightly different argument in the *Epinomis*, he observed that because stars don't stray or wander from their courses, intelligence existed in the regularity of their movement.[32] For Ptolemy, astronomy alone could know 'the attributes of those beings which are on the one hand perceptible, moving and being moved, but on the other hand eternal and unchanging'. Thus his job as an astronomer was a divine calling, where divinity is understood as a principle of order involving the motion of spherical bodies.[33] Ptolemy's ethics are often overlooked when discussing his mapping work, where concerns about the use of mathematical projections of latitude and longitude are more prominent. He was exceptional in the degree to which he emphasized astronomy and, therefore, mathematics as the ethical keys for unlocking the door that would bring him closer to the celestial divinities.[34] Ptolemy preferred to discuss the divinity of the heavenly bodies and their association with the classical gods separately from their mathematics. It was the job of astronomy to describe the regular movement of the heavens. Understanding their movements so that humans could arrange their affairs accordingly required a study of the harmony between them, an ethical undertaking. Ptolemy likened the harmonies in the movements of the stars to a 'chain of hands joined in a circle in a dance, or like the circle of men in a tournament who assist each other and join forces without colliding so as not to be a mutual hindrance'.[35]

Only by understanding the celestial and ecliptic poles can one arrive at an understanding of the Earth's North Pole and its

In this exquisite 15th-century 'Tapestry of the Astrolabes', the Pole Star in the centre is surrounded by constellations in the form of an astrolabe, the most important medieval instrument of navigation. The astronomer Hipparchus and poet Virgil are shown on the right with Philosophy on the throne.

central importance to Ptolemy's *Geographia*, his great work on 'world cartography'. Its purpose was to show how to map the known inhabited world – the *oikoumene* – following the ethnic and spatial diversity of trade routes spanning Europe, Asia and Africa, to the extent that they had been measured and recorded. This meant showing places in their true proportions and locations using the mathematical measures of latitude (parallels) and longitude (meridians). Meridians are in fact great circles that meet or pass through the poles. Without poles, there can be no meridians. That is why Ptolemy explains, 'the first thing one has to investigate is the earth's shape, size, and position with respect to its surroundings [heavens].'[36] The system of latitude and longitude used for this purpose came from astronomy.

The geographical North Pole also played a key role in the second stage of making a world map, the determination of the countries comprising it, in their correct place in relation to the globe as a whole. The longitude of places depended on lunar eclipses being observed at two locations and such astronomical sightings were rare. Latitude could be measured more easily by the height of the Pole Star. Ptolemy complains that over the previous three centuries, 'Hipparchus alone has transmitted . . . elevations of the north pole of a few cities . . . and [lists of] the [localities] that are situated on the same parallels.'[37]

The geographical North Pole, where all meridians meet, lay beneath the celestial pole in the heavens. In Ptolemy's cosmography, the celestial pole was on the axis of the universe around which all other heavenly bodies rotated and was in that sense real, even though there was no Pole Star present nearby. The Earth's North Pole was also located on this celestial axis, with all meridians meeting there. In Ptolemy's universe, however, the Earth was taken to be still and not rotating, and in that sense the Earth had poles but no polar axis! So for Ptolemy, the celestial North Pole was a point of real significance in the universe, whereas the geographical North Pole was neither more nor less than a mathematical point on the axis of the universe, but whose importance was chiefly to enable the making of two roughly contemporary developments, a world map or a globe.

The North Pole was also implicated in creating a different kind of cartographic space – the empty space of Ocean beyond the *oikoumene* whose limits Ptolemy had assigned to Thule at 63°N. Greek astronomers coined a term, rarely used, for the realm or region between the latitudes of the Arctic Circle and the North Pole itself – the *periskian* – referring to the edge of the world. Ptolemy divided the Earth's surface into climatic zones (*klima*) based on the length of the longest days of the year. Between the Arctic Circle and the geographical pole, there is a time of year when the shadow cast by a sundial would in theory complete a full 24-hour circle – the *periskian*.[38] Had Ptolemy or one of his travellers spoken to an Inuit hunter using his harpoon planting a gnomon in the snow or icy ground, he would have gathered a great deal of useful information, but based on the available evidence, Ptolemy believed the entirety of the *periskian* to be uninhabited.

Throughout classical Greek thought there remained a tension between a view of the world from the *oikoumene*, looking out, and a liminal northerly world on the outside or the edge of the world, looking in. This was much more explicit in Greek mythology, where the Hyperborean people of the North under the leadership of a prophet named Olēn are said to have founded the oracle of Apollo in Delphi.[39] This movement from the northern outer edge of the world to its inner sanctum is said to represent 'the signatures of exogenous rather than indigenous heroes . . . marking a spot where mysterious outsiders have come and gone'.[40] Hence the polar Hyperboreans not only look in at ancient Greece from the outside; they actually play a formative role in its conception. This north to south movement sets the scene for one of the Greeks' own endogenous gods, Herakles, to travel out to the ends of the Earth to demonstrate that his authority surpasses and tames all liminal strange or monstrous powers like the Hyperboreans on the margins of imperial authority. In ruling the outer margins of the known world, he declares his authority over the entire world.

Pytheas' readers, learning that Thule was situated at about 63°N, assumed that the world further north was uninhabited and unimportant. The migration of the Inuit's predecessors up into northern Greenland extended human knowledge of the Arctic another ten to fifteen degrees of latitude, but access to food in these very high latitudes could be precarious and oral histories reveal that the possibility of periods of famine was real.

For the Inuit and Micronesians, the Pole Star (Polaris) would come to play a role in celestial knowledge rich in metaphysical or mythical meaning. The irony, of course, was that the altitude of the Pole Star meant that its practical significance largely ceased for people living north of about the 70° parallel. For the Greeks, however, the celestial and ecliptic poles were key reference points for constructing star catalogues, the most famous of which was Ptolemy's *Almagest*.

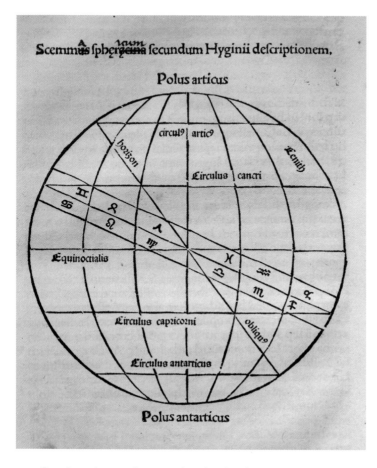

Scemmus sphçrçtna secundum Hyginii descriptionem.

Polus articus

circul⁹ | artic⁹

Circulus | cancri

horizon

zenith

Equinoctialis

Circulus capricorni

obliqus

Circulus antarticus

Polus antarticus

In the astronomy of Hyginus, probably drawing on Ptolemy's *Almagest,* the celestial Arctic or North Pole is part of the basic infrastructure of the heavens. Published in the *Poeticon astronomicon* (1485) in Venice, it contributed to the rediscovery of ancient astronomy.

On the subject of geographical poles, however, Ptolemy had little to say. For him the geographical poles occupied a relatively unelaborated space, external to the known or observable world, beyond the pale of the inhabited community or *oikoumene.* The geographical poles in Ptolemy's world geography played a role in defining longitude, holding all the great circle meridians in place. This framework made possible the construction of world maps and globes, and therefore the mapping of the *oikoumene* in its correct proportions and size. Astronomy would in the long run continue to trump geography as a source of scientific truth and as a principle for orienting ourselves on Earth. The geographical North Pole would, however, take on real importance in early modern Europe, becoming one of the most visible, beautiful and prized emblems of cartography. How maps with geographical poles taking centre stage came to be so popular in early modern Europe is taken up in the next chapter.

## 2 Holding the North Pole

The origin and spread of beautiful polar maps in the early modern period presents us with a paradox. How could it be that people were able to see the geographical North Pole without actually being near the North Pole? This riddle holds the key to unlocking the tradition of polar vision, that is to say the ability to imagine looking down on the Earth from above the North Pole in a way that, by the time the Royal Society was founded in 1660, had become commonplace among ordinary educated people in Europe. Centuries before being visited, the North Pole could be grasped, held in one's hands and described with images and words – without crossing the Arctic Circle. The answer to the riddle is that in the early sixteenth century a group of Renaissance artisans began to construct and popularize the North Pole in order to show the cosmographical relationship between the heavens and the Earth. New kinds of maps, globes and other instruments meant that the North Pole could be held and gazed upon by those with an interest in doing so. The remarkable way in which this happened in many ways laid the foundation for our polar geographical imagination right up to the present day.

Peter Apian (1495–1552), born into a well-to-do, middle-class family of shoemakers in landlocked Saxony, might seem as unlikely an individual as one could hope to find to begin a story of polar mapping. Although he had no appetite to become a mariner or explorer, he was nevertheless ambitious and talented in his love of cosmography. As we shall see further on, he became one of the most important popularizers of cosmographical works

Peter Apian
(1492–1552) was
part of a circle of
leading European
mathematicians,
cosmographers and
publishers who made
polar and other
spherical projections
an integral part
of early modern
map-making.

PETRVS APIANVS LEISNICENSIS,
*Divi Imp. Caroli V. Mathematicus et Comes.*
*Palat. Caes. Equestr. dignit. et in Academia Ingol.*
*stadiana Mathes. Profess. Publ.*
*Nat. A.* cɪɔ ccc xcv.  *Denat: A.* cɪɔ ɪɔ lɪɪ.

in his day and would publish the sixteenth-century equivalent of
a best-seller. The print revolution that had transformed Europe
meant that he was able to publish the North Pole in a form and
for an audience that previous generations would scarcely have
imagined possible. After studying astronomy and mathematics
at the universities of Leipzig and Vienna, he saw that a practical
calling could be no less exciting and set up a small print workshop. For many years he dedicated himself to exploring printing
and developing innovative techniques for bringing cosmography
to life for the book market. One key skill he worked hard to
master was to make complex astronomical principles digestible,

understandable and interesting to lay readers with a curiosity about worldly matters: merchants, princes and tutors.[1]

In the sixteenth century the study of cosmography meant paying close attention to the harmonies governing the relationships between the Earth and the constellations, stars and planets in the heavens. The timing of these movements of heaven and Earth were of great importance to astrologers and their wealthy patrons – the emperors, kings and queens of Europe – for the heavens were an instrument for avoiding or, if need be, managing crises and cataclysms. Apian realized, in ways that others were slow to grasp, that political and commercial elites could not manage without the North Pole. This distant and unknown place on the surface of the globe was not necessarily interesting in its own right, but Apian worked out that it was indispensable for making cosmography understandable to an audience still struggling to come to grips with the sudden expansion of geographical knowledge in the wake of the voyages of Christopher Columbus (1451–1506), Vasco da Gama (c. 1460–1524) and Ferdinand Magellan (1480–1521).

World maps failing to show the routes to the spice trade in the East Indies, or to the plantations in South America and the Caribbean, were soon hopelessly out of date. For gentlemen, an overview of new discoveries on the other side of the world, while containing them within the perfect cosmographical shape of a circle, was a sign of their own expanded horizons and social status.[2] Patrons whose commercial interests increasingly spanned the globe turned to cosmographers and map-makers for new ways to visualize the globe, to understand its principles and to know its character. The historian Jerry Brotton has rightly referred to this moment in European history as the advent of a new way of envisioning a globalized world, what he terms 'terrestrial globalism'.[3]

Apian established his printing business more than a decade before the Copernican revolution of the 1530s. Students of cosmography were then being taught Ptolemy's model of the universe with the Earth at its centre. Apian, who had the mathematical ability to interpret Ptolemy's writings on astronomy and geography, spotted an opportunity to meet the growing demand from tutors and patrons who wanted to be able to understand how the Sun and planets moved harmoniously around the Earth. For the frontispiece of his *Cosmographicus liber* (1524), Apian produced a woodcut of a globe showing Africa and Asia, not the familiar Europe of old. To introduce his readers to the place of the Earth, its meridians, tropics and poles in the structure of the universe, he showed them in their positions on celestial rings. The alignment of the Earth's polar axis with the celestial Arctic and Antarctic poles was plain to see. Readers with any grounding in cosmography were familiar with this fundamental principle. They knew that the Earth itself stayed absolutely still as the spherical shells holding the planets and stars rotated around the Earth evenly and constantly. That itself was hardly new, since the astronomical theory that Renaissance astronomers and cosmographers were working with had been inherited or rediscovered from the ancient Greeks. This was possible thanks chiefly to the work of Arab astronomers, who had not only kept Ptolemy's ideas alive, but had developed them through centuries of study and the invention of ingenious astronomical instruments. Thus Apian and his contemporaries didn't invent the polar orientation of globes and charts out of thin air: it emerged in Europe out of the encounter and interplay of ideas moving between medieval and early modern European and Arab civilizations.[4]

When we think of popularizing science today, we immediately think of simple, clear prose and pleasing illustrations. Apian used prose sparingly; he understood that ordinary people could more easily visualize the Earth's place in the universe through diagrams and models, rather than relying too much on verbal explanations or mathematical formulae. Being a mathematician himself, he worked hard to convey what he understood to readers, many of whom had minimal training in mathematics. As a result his popular books were much sought after and, with such demand, could be priced affordably. Even after he achieved popular commercial success and turned his attention to catering for the taste of his patron, the emperor Charles v, the visual dimension remained prominent, and even dominant. The striking feature

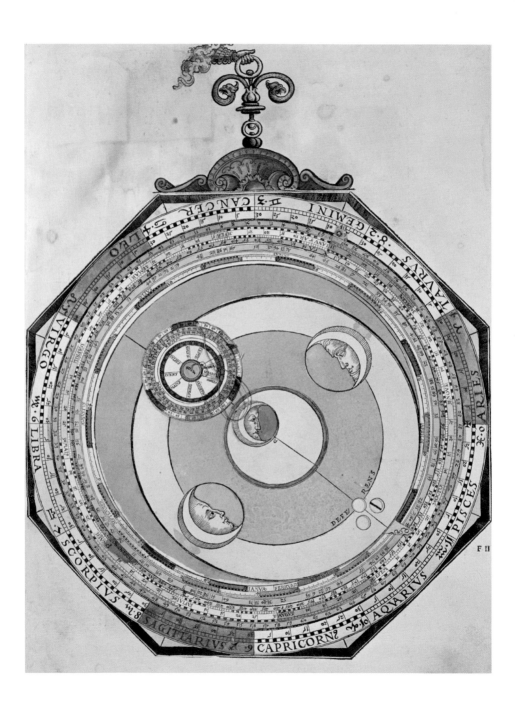

of Apian's magnum opus – his extravagant baroque *Astronomicum Caesareum* (1540) – was the exquisite construction and iconography of a set of moving astronomical dials.[5] As a printer of cosmography, Apian's dedication to visualization meant being a craftsman. By holding open his books, one was more than a reader of words; readers could use these remarkable paper instruments to participate in cosmography by aligning or orienting themselves with the heavens.

Apian's real polar innovation for many readers of *Cosmographicus liber* was to allow them to view the new map of the expanded world by imagining themselves looking down over the globe from above the North Pole.[6] To be able to picture a place was halfway to knowing it. Readers with feet firmly on the ground might be accustomed to viewing the night sky, but picturing the reverse, looking down from above, involved a kind of abstraction that was altogether more complex. In keeping with the spirit of cosmography, they needed to learn to align themselves with the universe's divine symmetry; they needed to have been shown a celestial globe to see that the heavens were spherical; and they then needed to see that the Earth could be viewed from a distance at a time when perspectival drawing was still a relatively new idea. This way of seeing the Earth stemmed from what the geographer Denis Cosgrove calls the 'Apollonian gaze', after the Greek god who could see the entire Earth by flying over it.

The cosmographers of Apian's time, particularly those working in universities or at the royal court in Vienna, developed a small number of 'polar' or 'circumpolar' projections, maps of symmetric concentric circles around a central point, the North Pole, from which straight lines of longitude radiate outwards. Then, as now, polar maps were appreciated for their symmetry and beauty. Within a hundred years they would be associated with maritime navigation and were often used as the frontispiece in world atlases. These special polar maps conferred beauty and prestige on atlases in a unique way. These projections acquired an aesthetic significance and a prominence that became synonymous with a new way of viewing the world, as though gazing down on the world from the celestial pole.

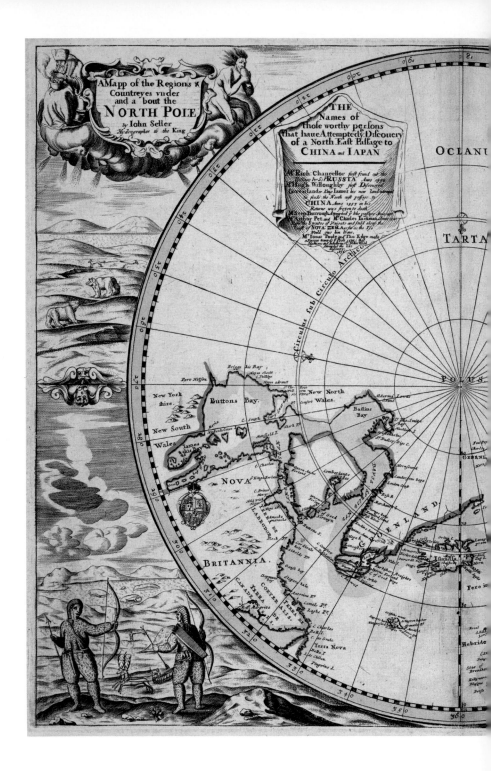

A Mapp of the Regions &
Countreyes vnder
and a bout the
NORTH POLE
by Iohn Seller
Hydrographer to the King

THE
Names of
those worthy persons
That haue Attemptedly Discouery
of a North East Passage to
CHINA and IAPAN

Mr Rich: Chancellor first found out the
Passage by Sr RUSSIA Anno 1566
Mr Hugh Willoughby first Discouery
Greenelandor Kinge James his new Land attempt
to finde the North east passage to
CHINA Anno 1553 in his
Returne was frozen to death
Mr Steph Burrough Attempted & this passage Anno 1557
Mr Arthur Pet and Mr Charles Iackman Anno 1580
the Streights of Waigats and saild along the
Coast of NOVA ZEMLA as far as the Ise
Would cut him home.
Mr Ionas Poole and Tho: Edge made
a Farious view of Edge, Cherie Isle.
Hackluts and Herfords and Mills Iled
Attempted the like.

OCLANI

TARTA

POLVS

Circulus sub Circulo Arcticon

Briggs his Bay
Noys chistt
C. Pellipe
Noys ahroitt

Perre Nelson

New York
shire

New North
Wales

Buttons Bay.

Baffins
Bay

New South
Wales

James
his Bay

NOVA

GEENL

NORD

GROENLAND

BRITANNIA.

Terra Nova

Pilgrims I.

Hebride

The scenes of hunting and whaling framing John Seller's polar 'Mapp of the Regions and Countreyes under and about the North Pole' (*c.* 1676) convey a view of the Arctic as a prosperous region of commerce and empire.

To make sense of the rise of polar projections requires taking a step back to see why organizing space using circles of latitude and straight radial lines of longitude came about. Latitude–longitude is now so familiar as to seem self-evident, as though it were the only universal or scientific system for specifying locations on a globe with accuracy. But is it really the all-encompassing natural way to define positions? After all, many non-Western cultures got along just fine for millennia without this. For Renaissance cosmographers, the system of latitude and longitude was uniquely valuable because of its utility to navigators and astronomers. Locating the positions of stars and planets required measuring their elevations from the horizon or their distance from other bodies. Poles, of course, were mathematically defined points used to project a grid of latitude and longitude onto a celestial sphere. Ancient Greek astronomers like Ptolemy had shown that this grid could also be applied to a world map of the Earth. Lines of longitude cannot be generated without poles to pass through: without poles, there can be no longitude. In the celestial realm, the position of stars was observed using angles measured in degrees, minutes and seconds in relation to their latitude and longitude.

Today it is sometimes said that the North Pole is only an imaginary point, a mere mathematical construction. This is true, strictly speaking, in the sense that poles are purely abstract points used to project time and space, and are at the heart of Ptolemaic cosmography. But it is misleading to think that because poles are zero points in geometrical projections, they are therefore immaterial. Cosmographers like Apian not only trafficked in abstract mathematics; they worked closely with some of the most highly skilled artisans of their age: painters, engravers, paper makers, ink makers, bookbinders, leather craftsmen and the list goes on. Mathematics could be very practical when working hand in hand with artisans; in fact the term 'practical mathematics' became an important branch of study in this period because of its role in making instruments for surveying and drawing. Through collaborations in European workshops throughout the sixteenth century, the North Pole – on Earth and its celestial counterpart – came alive. Thanks to these artisans, the poles became something visible and tangible, something one could see and lay one's hands on.

The story of how the terrestrial North Pole came to matter at the dawn of modernity has remained unexplained because other projections for displaying the world on a map were better suited to showing tropical and temperate zone trade routes. Apian enjoyed making polar maps: they were to some extent cartographic novelties, objects of beauty to admire. However, they also played a unique role in the design and construction of a range of cosmographical instruments. They had the property of aligning the eye with the axis running through the celestial and geographical poles. This turned out to be especially useful for enabling readers to observe celestial relationships such as the movement of the Sun across the sky or the position of fixed stars at different latitudes. It enabled readers to position themselves as observers not just in London, Paris or Vienna, but on a line of latitude circling the entire globe. This global perspective meant that the North Pole played a very special role in the construction of these cosmographical instruments. They required a pole to act as a focal point, holding the different material parts of the instrument in place, in order to coordinate the movement of the heavens and Earth. As we shall see using several examples, it is no exaggeration to say that the poles held the globe together.

## Cosmographers' polar projections

When Apian decided to publish polar maps of the heavens or Earth, he needed a polar projection. If polar maps of the stars were going to be at all useful, the projection onto the flat paper surface would need to preserve the elevations and other angles of the stars observed by astronomers. Tackling this was a mathematical problem that required techniques for working with angles on spherical surfaces. Apian was fortunate to have studied in nearby Vienna, introducing him to the work of a circle of highly talented mathematicians in Nuremberg, Ingolstadt and

Vienna who were working under the patronage of Maximilian I, Holy Roman Emperor (1459–1519), who had arranged marriages for his children and grandchildren that helped bring much of Europe and half of the New World under Habsburg rule. Getting to grips with spherical angles was part of the new spherical trigonometry that rested on a critical engagement with Ptolemy's mathematics, astronomy and geography.

Johannes Stabius (1450–1522) and Johannes Werner (1468–1522) applied this mathematics to Ptolemy's geography of the Earth to introduce new projections. Stabius, a member of the circle of Vienna humanists, had developed the mathematics for the

Albrecht Dürer, *Emperor Maximilian I* (1459–1519), Holy Roman Emperor, *c.* 1519, woodcut.

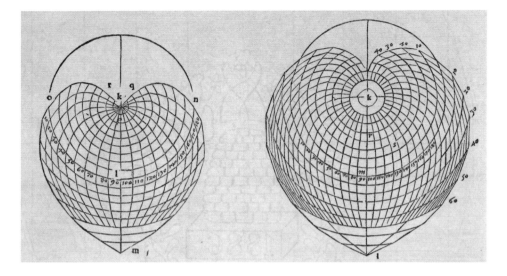

Johannes Werner's drawing from his 1514 commentary on Ptolemy. The association of piety with the cordiform map projection's polar-focused heart shape increased its appreciation among cosmographers.

heart-shaped or cordiform projection (*c.* 1500), which, though part of the family of conic projections, focuses the eye on the North Pole. Werner, specializing in spherical trigonometry, developed the cordiform projection as well as working out the mathematics for the polar planisphere projection. By dedicating himself to the study of Ptolemy's *Geographia*, and publishing a scholarly Latin translation of it in 1514, Werner was able to utilize his mathematical learning to present Ptolemy's work to an audience in a new age.[7] This brought new visual prominence to the North Pole and, emanating from it, the angular divisions laid down in meridians or lines of longitude. These cartographic projections were adopted by cosmographers and cartographers, including Gemma Frisius, Gerardus Mercator, Oronce Finé and Abraham Ortelius.

The cordiform and planisphere projections spoke to the needs of Maximilian I and his grandson Charles V (1500–1556), who brought together the Austrian and Spanish Habsburg possessions. They staked their power on claims of universal authority to preside over an empire circumscribing the globe. Displaying this authority called for artisans and instrument makers to innovate and put these new mathematical projections to work. Globes and charts were relatively new and rapidly changing technologies in this age of European empires. They were a beautiful means of displaying extraordinary power and picturing a world that in a generation had burst the bounds of medieval imagination. Terrestrial globes, of little interest a century earlier, were becoming grand objects in their own right, deserving a place in the libraries and salons of princes, often constructed as pairs with celestial globes. Hans Holbein's painting *The Ambassadors* (1533), showing a small terrestrial globe accompanying a much larger celestial counterpart, was one example of the growing prominence of terrestrial globes.

The projections of Stabius and Werner found a niche in the market for readers in the merchant classes. Globes were generally very expensive, upmarket objects that were time-consuming to build, reflecting their status along with maps as a form of mathematical instrument (a distinction not usually recognized today). The great popularizers Apian and Gemma Frisius recognized the value of creating less expensive instruments that could take advantage of the printing press without the need for the expert and painstaking work of teams of artisans. Beyond the small circles

Experiments with map projections enabled cosmographers like Bernardi Sylvani (1511) to present the ancient geography of Ptolemy in new ways for audiences keen to understand their rapidly changing world.

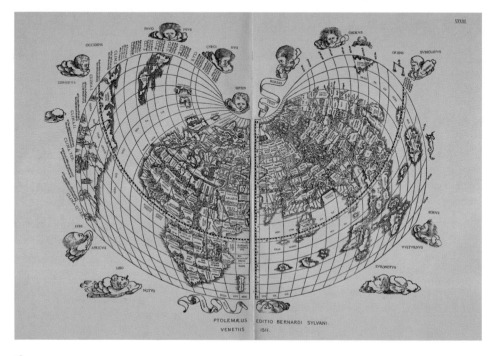

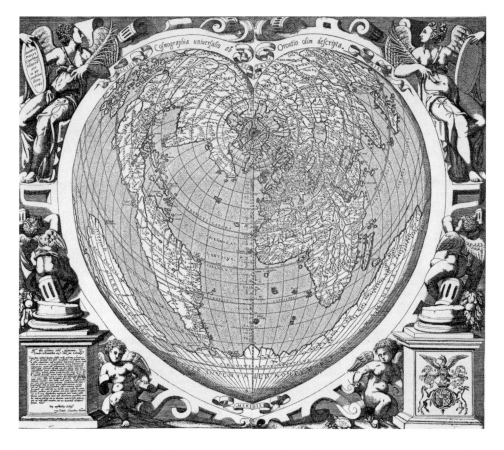

Oronce Finé, making use of Werner's second projection, produced an impressive cordiform map of the world with 360° coverage around the North Pole (copper-print edition by Cimerlinus, 1566, reproduced by Nordenskiöld, 1889).

of erudite astronomers and mathematicians, a prosperous class of Renaissance readers, educated in Latin and anxious to understand their rapidly changing world, were willing purchasers of new forms of knowledge. Instruments resembling pop-up charts could appeal to teachers, mariners and readers with an inclination for cosmography.

Among this proliferation of relatively inexpensive mathematical instruments, constructed mainly out of paper, one finds, most unexpectedly, polar maps. Why is this so? Polar maps combined a new geopolitical perspective with an appeal to universal authority whose power reached beyond the self-interest of nations. This remarkable array of polar instruments helped to make sense of the new riches being returned through trade,

59

colonization and piracy on the other side of the world. These exciting new world maps and charts were keen to show the most recent discoveries in the Americas and the East Indies. Spanish, Dutch and Portuguese merchants who sailed into the Indian and Pacific Oceans were transforming world geography. The classical framework of Ptolemy's world, defined around the *oikoumene* – the inhabited world of Europe and northern Africa centred around the Mediterranean – was buckling under the pressure of the new knowledge streaming in from these empires. New tools for envisioning the world were in demand by people with power and money.

Where the prestige of powerful imperial patrons was at stake, cosmography and cartography accompanied new ways of representing power. Charles V expected much in return for his patronage of astronomers, cosmographers and geographers. They in turn understood the importance of depicting the emperor as a modern-day Apollo. Doing so placed him in a lineage of imperial rulers dating back to Alexander the Great, himself the patron of Aristotle. This iconography showed for all to see that Charles V, like Apollo, could cast his gaze over the globe in its entirety. This aerial perspective, the hallmark of claims to universal authority, was in reality bestowed on emperors by artisans as well as the courtiers who designed their rituals, heraldry and other symbols.

Introducing the perspective of viewing the Earth from above brought cosmography into line with the new developments in drawing, projection and perspective pioneered in Renaissance Europe. Albrecht Dürer (1471–1528), one of the most remarkable German artists, was the son of a prominent goldsmith in Nuremberg. Dürer's precocious talent for drawing broadened into printmaking, writing and an extraordinarily rich span of philosophical interests. His studies of perspective spanned much of his life and he brought back to northern Europe the principles of linear perspective he encountered while studying in Bologna. He later moved to Vienna to work with Stabius and Werner under the patronage of Maximilian I. Dürer and Stabius published the first polar star chart in 1515. He is also known to have prepared an unpublished polar terrestrial chart for which there

Polar projections of the Earth acquired importance because they conformed to the symmetry of the celestial realm and the universe. Albrecht Dürer's 1515 chart of stars and constellations is oriented around the ecliptic pole, the centre of the plane of the Sun's orbit.

was as yet no substantial market.[8] Johannes Werner and Albrecht Dürer were both from Nuremberg and formed a close friendship. Both drew inspiration for their spatial mathematics from the Italian Renaissance, particularly the work of the Florentine engineer-architect Brunelleschi on the development of linear perspective.

Werner and Dürer were themselves closely studied by the next generation of cosmographers, notably Peter Apian and Gemma Frisius, who explicitly acknowledged these debts. In his

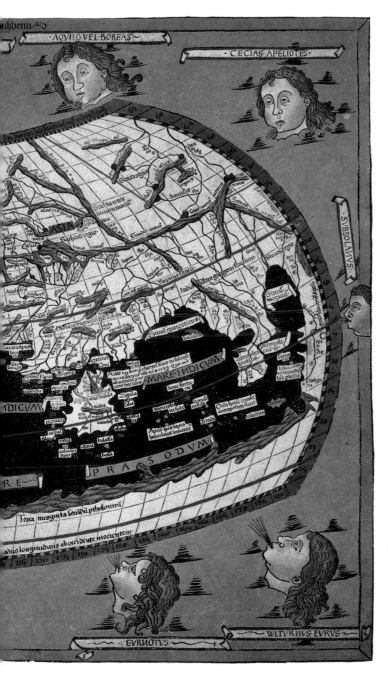

Ptolemy's map of the world, oriented around the Mediterranean, was published in a Latin edition in 1482. Though approximating a section of a sphere, the known world largely comprised Europe, Asia and Africa.

*Cosmographicus liber*, the work that made Apian a household
name, one illustration using linear perspective depicts an eye
external to the world viewing the Earth and heavens from a
distance. This may be one of the first cosmographical diagrams
to show explicitly how a human eye might observe an earthly
pole from a distance. Whereas astronomers and navigators,
rooted to the surface of the Earth, had gazed upwards at the

fixed celestial pole, Apian was teaching readers how they might also mirror this upward gaze, looking at the Earth depicted with human beings walking on the surface against a backdrop of the world around it.

One can see in Apian's image that he is inviting readers to reflect on how the eye is aligned with terrestrial and celestial bodies, just as observers were doing when using mathematical instruments for observation. Learning to use an instrument required observers to understand how to position themselves to hold an instrument correctly, so that being properly aligned, a scale on the instrument could then be read to give a measurement expressing the relationship between the observer and objects in the cosmos. Learning to see in this way was part of working with a wide range of practical instruments.

Depicting polar vision in terms of the geometry of the eye had a precedent because some existing instruments required observers to align their vision with the celestial pole.[9] The astrolabe, for example, was an instrument that enabled observers to show the position of the stars in the past or the present, and in navigation to find one's position, direction and local time. To hold an astrolabe required using one's hands to align the two sights with the celestial pole. The nature of most practical instruments was that they had moving parts so as to be able to coordinate the relationship between different celestial bodies like the Sun and the (other) stars. On the astrolabe the polar axis coincided with its centre pin, which had the function of coordinating the moving parts of the instrument to track the pathways of celestial bodies.

Newly invented navigational instruments like backstaffs and nocturnals also required their users to align the instrument's sight with the Pole Star. A backstaff would give a mariner his latitude simply by measuring the height (angle from the horizontal) of the Pole Star. In the absence of a Pole Star, the nocturnal was invented to track the path of the nearest star around the celestial pole, which could then be used to correct the backstaff's error. Aligning the eye with the celestial pole was therefore already a well-established practice of observation.[10]

Aligning an observer's eye with the celestial pole was only a short step away from aligning the eye with the geographical North Pole. Projections could be thought of as instruments for positioning the eye. The planisphere projection, later called the polar stereographic projection, was precisely such an instrument for aligning the eye so as to look backwards towards the Earth. Its rudiments had been known to Hipparchus and Ptolemy. The mathematics of Stabius and Werner had solved the problem of spherical angles so that, when using this projection, angular proportions or measures were correctly preserved.

The first decades of the sixteenth century were crucial for turning the planisphere into what would become a standard cartographic object. Early examples by Gualterius Lud (1507), Gregor Reisch (1512) and Dürer (c. 1515) point to the sense of innovation in making polar maps that presented the whole world, using means that enabled both eastern and western hemispheres

This simple woodcut showing a navigator using a backstaff to find his position with the Sun's altitude illustrates how navigation and astronomy taught observers to align their sight, and hence their bodies, with the heavens.

Apian introduces his readers to the baisic architecture of cosmography, displaying the orientation of the globe in relation to the celestial poles, the meridians, and the plane of the sun.

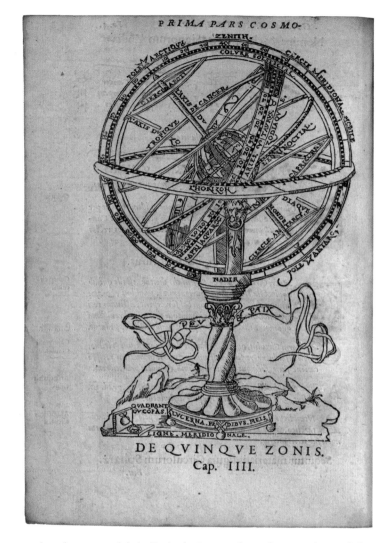

to be shown and labelled. Apian explained to readers of the *Cosmographicus liber* that, when using the planisphere, the eye is positioned in line with the axis of the universe and the celestial pole. Viewed externally from a distance, the meridians on a map resemble a star shape with its straight lines emanating from the terrestrial North Pole. Closely related were oblique versions (where the eye is offset at an angle from the celestial pole), which Werner classed among the four 'new' projections included in his edition of Ptolemy's *Geographia* (1514). Werner offered a valuable hint of the marketing value of these projections when he recommended the planisphere as being well suited to 'men of distinction' wishing to have 'an ideal representation of the globe'.[11]

## The terrestrial polar speculum of Apian and Gemma

In Apian's day, Vienna and Louvain (now in Belgium) were both centres of learning in the Holy Roman Empire. Charles v came to rule over the combined European territories of Spain, Burgundy and the Habsburgs, as well as the conquered territories of the Spanish New World. Intellectual and financial opportunities were there to be found in publishing for less elevated audiences as well as for elite patrons. The biographies of Apian and Gemma became intertwined by their shared passion for making cosmographical instruments. Both Apian and Gemma had moved from small provincial towns to university towns, where academic learning and freedom of thought flourished more easily than in religious centres. Whereas Apian had moved to Vienna and then Ingolstadt to study mathematics, Gemma had moved from Friesland to Louvain to study medicine while devoting every available moment to cosmography. Apian and Gemma were eventually able to support their love of making cosmographical instruments by holding salaried teaching positions at universities.

What binds Apian and Gemma for our purpose was the *Cosmographicus liber*. Started by Apian, it was much expanded and improved by Gemma, who republished it five years later, replacing Apian's woodcuts with engravings and using a less heavy font, which improved its appearance, and adding discussions and appendices. Intellectual borrowing was complex and it wasn't unusual for authors to borrow freely or simply take the work of others. Editions of Gemma's *Cosmographicus* contributed to the reputation of both authors. He in turn became the teacher and mentor of Gerardus Mercator, one of the most celebrated cartographers of the age, and John Dee (1527–1608), who would become a magus in the English court of Elizabeth I.[12]

Peter Apian's *Speculum Cosmographicum*, an instrument of moving parts constructed from paper, popularized the polar stereographic projection throughout Europe. The speculum, meaning 'mirror', gave readers a way of seeing their place in the universe by showing the movement of the Earth through the constellations of the zodiac. This early world map also shows the northern hemisphere of east (*oriens*) and west (*occidens*), 1551 edition.

*Overleaf:* Mercator's Pole-centred map of 1595 showing a large black rock (*rupes nigra*) at the North Pole, a magnetic mountain, surrounded by a whirlpool fed by four rivers dividing a continental landmass. Drawing on sources as early as Ptolemy, Mercator's geography illustrates how the speculation surrounding the pole was integral to understanding the globe, and not simply fanciful.

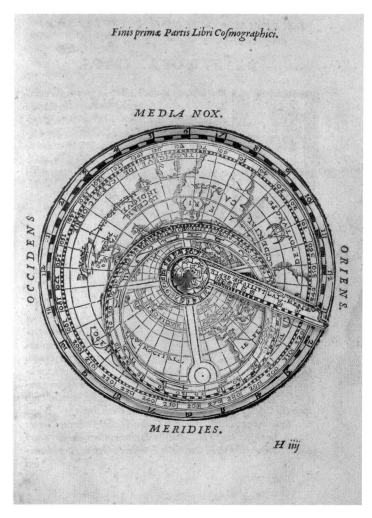

This gives one a sense of the intellectual legacy of this lineage of sixteenth-century cosmographers.

Apian used his new publishing business to learn, borrow and experiment with making popular educational instruments. He was skilful in exploring the market for his books, tapping into a readership wanting to get to grips with discoveries in the New World and understand how astronomy and cosmography shed light on the Earth's place in the universe. Apian therefore went out of his way to emphasize the importance of terrestrial

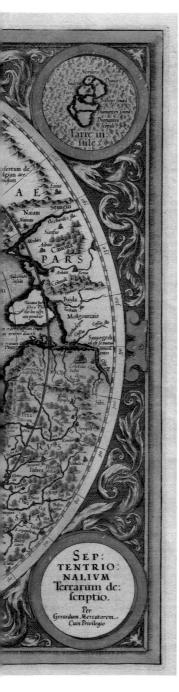

or earthly perspectives in cosmography: the geography of the whole Earth, the techniques of surveying and the view of the heavens seen with an upward gaze. He was grounding a terrestrial perspective in the harmonious certainty found in the regular movements of the celestial spheres.

Polar maps featured prominently in Apian's production of cosmographical instruments. The influence of Stabius, Werner and Dürer was plain to see in his inventiveness using polar views to communicate a cosmographical view of the Earth: first a polar world map (1520); soon after a circular world map with south at the top and only six place names (1521); and then the remarkable *Speculum cosmographicum*, meaning literally, a 'cosmographical mirror' (1524). This was a volvelle, a paper instrument with moving parts, bound and published in the *Cosmographicus liber*.[13] The cosmographical mirror employed a terrestrial polar map as its *mater* or principal plate. As the name implied, it enabled readers to reverse their gaze, and rather than orienting themselves upwards towards the celestial pole, they could gaze down from above upon the world, with the geographical North Pole at its centre.

Within ten years, Apian was also publishing a pole-centred, heart-shaped, cordiform map (like Oronce Finé, who was inspired by Werner). In his *Horoscopion Apiani* (1533) he included a celestial polar view map showing Arab and Bedouin constellations, in which he reversed the astronomer's normal view, looking down from outside the heavens, and instead showed the terrestrial vantage point of someone on Earth gazing upwards at the constellations of the zodiac. The sheer range of printed polar views provides a really important indication of their impact on how the globe was being imagined in this age of European expansion. To explain how readers thought about the Earth's North Pole, however, it is necessary to look more closely at what Apian hoped his readers would do with these instruments.[14]

Volvelles reached the height of their popularity in the sixteenth century with the rise and fall of cosmography.[15] Readers could use them to trace the steps in solving relatively simple cosmographical problems that had until then been restricted to learned astronomers, mathematicians and the most accomplished of navigators. These books were relatively inexpensive for publishers to produce. Paper was comparatively cheap, and with the advent of engraving, printing plates in large numbers became faster and easier.

Crucially, readers were invited to take on the work of assembling the instruments themselves, which simultaneously involved them in making the book and saved the publisher both time and expense.[16] It was in this context that we can see how successive editions of the *Cosmographicus liber* encouraged several generations of readers to invest time not only in viewing circumpolar maps in books, but in actively using them to construct simple polar instruments.

The *mater* or principal plate in the speculum was engraved with a stereographic polar map of the terrestrial world, centred on the geographic North Pole. On the *mater*, all meridians projected outwards across the globe. Mounted on the *mater*, and pivoting around the North Pole at the map's centre, was a rotating paper ring called the *rete* (akin to an astrolabe), which traced out the changing path of the Sun along its ecliptic plane at 23½° to the plane of the equator, and finally a rotating index arm showing the latitude of anywhere on the terrestrial map.

Apian explained to his readers how the Sun follows a path orbiting its own ecliptic pole, offset from the celestial pole, so that they could follow the Sun's position overhead when viewed sweeping across the Earth, and see how this changes throughout the year. A reader capable of using this instrument gained the practical reward of being able to predict the Sun's path on any date and time of day. Similarly, watching the Sun's movement around the terrestrial North Pole and across the face of the Earth allowed readers to see more clearly the difference in time between different places on the globe (this was prior to time zones).[17] This would also lay the ground for understanding more complex phenomena like eclipses and other celestial conjunctions important for astrologers and astronomers. In this way the predictable and regular path of the Sun and stars around the globe could be squared with the earthbound viewer's experience of their seasonally changing paths.

The speculum's polar map mirrored contemporaries' understanding of contemporary geopolitics. The position and outline of the continents America, Europe, Africa and Asia were clearly delineated. Extending radially outwards and south-facing, the outer perimeter of the map reached the Tropic of Capricorn (23½°s). This encompassed most of the inhabited world known to Europeans, limits being challenged annually through Portuguese and Spanish voyages in search of new trade routes and imperial possessions. Readers of this (almost) world map could recognize the new imperial dimensions of their sixteenth-century world. Apian, using a pole-centred projection, was able to bring together in a single view the political mythology of a world divided into East and West. The two margins of the speculum's polar terrestrial map boldly announced this division: 'ORIENS' on the one side and 'OCCIDENS' on the other, demarcating the eastern and western hemispheres.[18]

The spatial reach of the emperor's sovereignty shown in cosmographical instruments was also present in other imperial emblems. Charles v's coat of arms was inscribed with the Pillars of Herakles. His motto was *plus ultra* – ever further. This imperial vision projected not only spatially outwards, but temporally into the past. Alexander the Great, in a classic act of self-mythologizing, had adopted the mantle of Herakles to legitimate his claim to hold heavenly power over terrestrial domains stretching eastwards beyond the Persian empire to India. The iconography of Charles v, like Caesar Augustus and Alexander before him, united the Apollonian vision of flying to the northern edge of the globe with the imperial symbolism of Herakles and the conquest of the Orient. For that reason, the view looking down over the terrestrial North Pole from above was an imperial perspective, contributing to a European mythography about imperial power invested in the domination of the Orient through trade and conquest. For emperors such as Charles v, this unified view of the globe symbolized his claim to supreme universal power over worldly affairs.[19]

## Holding the world in place

Apian's readers participated in this world-making in a very practical way. Readers themselves were invited to assemble Apian's speculum from the pages of the book. First they would carefully set about cutting out the shapes from the plates, and then build the instrument by placing the shapes in their correct positions. One more simple and seemingly insignificant step was actually crucial. They had to fasten the plates to each other in their correct positions by sewing or tying them together at the point of rotation: the North Pole. Apian provided his readers with a simple piece of thread or string for the purpose. It wasn't a difficult task, but it needed to be done properly to hold the cosmos together and to enable the reader to observe how its parts moved harmoniously. That was the whole point of cosmography. If fastened too tightly, the parts could not move smoothly; too loosely, and the Sun would deviate clumsily from its true path and fail to deliver reliable and accurate calculations; and if misaligned, the result would be an unflattering and unharmonious cosmos.

The humble fastening device, rarely discussed by historians, was a crucial piece of cosmographical kit that explains the meaning of the North Pole for Renaissance readers of cosmography. To the question 'What does the circumpolar map in the speculum show at the North Pole?', the answer is literally a space or a hole. The North Pole is marked by an absence of meaning rather than any inscribed place name or other indication. What was at the North Pole – nobody knew! Instead, the hole in the pole was designed for a different kind of presence. Apian provided a piece of thread or string; other artisans used rivets or pegs to hold the cosmos together. The function of these simple pieces of material was to allow the parts to rotate smoothly around a pivot point. The movement of the Sun and stars across the Earth was coordinated at the North Pole. Small fasteners made from pieces of metal, wood or thread might hardly be the subject of conversation to the owner of such an instrument, provided they did their job and weren't in need of repair. Yet were such a device broken or stuck, an instrument could not pivot or rotate, and without rotating parts cosmographers were very restricted in their ability to show the Earth's place in the harmonious motions governing the cosmos. Without simple fastening devices, Apian would not have been able to produce the kind of affordable volvelle that contributed to the popularity of the *Cosmographicus liber* and made it an exceptional publishing success that went through many successive editions.

Pole-centred projections, by virtue of their iconography, were also taken up in expensive high-end instruments like globes. Globes, too, were fitted with pins at the North Pole extending outwards from their inner central pillars, enabling the globes to rotate freely. Like the speculum, the holes in the globe enabled them to rotate around a pivot point. Globe-makers, however, often went one step further by using the pins to add information or meaning to the North Pole. For example, a simple circular ring became a popular way of dividing the day into twenty-four hours to indicate how the movement of the Earth caused time to vary with longitude.

On larger globes, the fasteners at the poles opened up a new space for the display of gratitude above and beyond the polar gores or caps. What could be more important to an instrument maker's patron or client than due acknowledgement of their power and taste? As globes became more prized, the pin at the terrestrial North Pole carried more prestige. In exceptional commissions, very expensive symbolic objects like silver clocks of a considerable weight were mounted above the North Pole. These ancillary devices – time circles, clocks, sculptures – reveal the central importance of enabling owners to show the Earth from different vantage points. Much more than mere decoration, these were the devices by which Apollonian visions of empire were created and assembled in the artisan's workshop.

In terms of seeing the Arctic regions as desirable territory to possess, English and Dutch navigators such as Richard Chancellor (*c.* 1521–1556) and Willem Barentsz (*c.* 1550–1597) sailed in the name of Protestant monarchs, searching for an alternate maritime route to the Orient. One can imagine educated readers, noblemen and merchants viewing the fabled Northeast Passage on a globe: positioning themselves to advantage, aligning themselves with important meridians, holding the globe firmly and rotating it gracefully on its terrestrial poles. In the period of expanding European global empires, displaying the globe was more important than speculating about the substance of the poles in their own right. True, whether the North Pole would obstruct navigation or offer an ice-free route would come to matter a great deal. In the first instance, however, the North Pole's role was to hold the Earth at the centre of a universe governed by harmonious movements of stars and planets, and to display the rapidly changing relations of navigation, political power and empire.

The circular Pole Cap on this exquisite terrestrial globe, constructed by Emery Molyneux (1603), provided an important space for a lengthy and elegant dedication to his patron, William Sanderson. The Molyneux pair of globes were the first to be made in England.

What the North Pole was actually like and what made it unique would puzzle natural philosophers for a long time to come. In the next chapter, we will see how William Gilbert built experimental models of poles to solve the riddle of the compass and the Earth's magnetic poles. Yet these same investigations would make the poles more strange and paradoxical, not less.

# 3  The Multiplication of Poles

William Gilbert (1544–1603), natural philosopher and London-based physician, moved in elevated social circles. As personal physician to Elizabeth I and a select group of wealthy patrons, he considered himself to be a philosopher equal to any of his contemporaries in England.[1] The terrella, a novel instrument largely of Gilbert's own ingenious design, held within its spherical body the secret principle of the universe. No larger than a piece of fruit or an ordinary ball, the terrella contained within it two poles, which he likened to those of the Earth, naming them north and south. Like a small handheld globe stripped of its geography and surface decoration, it had none of the normal materials of globes: none of the wooden posts, papier mâché, gores, inks or glues that went into their construction. The terrella was instead fashioned into its spherical shape from deceptively unprepossessing chunks of commonplace lodestone material, which could be quarried in any number of countries. To Gilbert, this polar instrument wrought from rock was a beautiful object, as mysterious and powerful as it was simple.

The terrella could do extraordinary things that not even the most skilled navigators since Pytheas had accomplished. Stephen Pumfrey, historian of Gilbert, put it this way: He reasoned that 'the Earth must be a magnetic sphere like the terrella, but also that the terrella was a reliable laboratory model of the Earth.'[2] In other words, the laws governing the Earth's poles could be modelled by an easily constructed magnetic sphere. The terrella, having magnetic poles of the same substance as the Earth, was

Portrait of
William Gilbert.

79

A blacksmith in his forge is shown working a piece of metal on his anvil to produce a magnetized needle, in William Gilbert, *De Magnete* (1600).

like the Earth in miniature. Whatever patterns of magnetic behaviour one could learn about the terrella through experimentation ought to be equally true for the Earth. The implication was that the mysteries of the poles could be brought inside a laboratory.

From a magnetic point of view, a ship could also be modelled in miniature so as to navigate over the surface of the terrella: 'His surrogate ship and compass was the versorium, a needle possessing polar magnetic attraction.' Having these instruments to model the behaviour of a ship's compass following a magnetic bearing was the experimental equivalent of being able to go 'anywhere [on the terrella] fast – even to the polar regions'.[3] Moving the versorium over the surface of the terrella's polar regions enabled Gilbert to observe magnetism at work very directly, though understanding the hidden causes of its behaviour was not so easy. His boasting about the power of experiment was of course implicitly in contrast to the difficult experience of Europe's most long-suffering Arctic navigators, worthies like Richard Chancellor (*c.* 1521–1556), William Baffin (*c.* 1584–1622) and Willem Barentsz (*c.* 1550-1597), in their quests to find polar passes to the Orient. What elevated Gilbert's instrument above the globes, compasses, backstaffs and other practical instruments of his day was that the Earth-shaped terrella contained something very special: the source of polarity itself.

Gilbert was more than an ordinary physician of his day. He was curious about all manner of substances around him – human, animal, mineral. He was an experimenter, a restless observer and manipulator of materials, and a philosopher at heart. He conceived, designed and commissioned new instruments like the terrella. In his laboratory he used his needle-shaped versorium to measure magnetic attraction at locations all over his terrella. Armed with this model, he built 'a systematic observational database that was beyond the combined expertise of the English, Dutch, and Spanish navies'.[4] In this new 'magnetical philosophy', as he called it, his terrella was a real living Earth in its own right, and not just an expensive model like a globe, a depiction of the surface of the Earth like a map. The behaviour of the versorium, deflecting towards the terrella's pole, was governed by the same principle that caused a mariner's compass needle to point northwards, but on a scale that could be mastered in a laboratory.[5] Beyond measuring the power of mutual attraction between pole and needle, what Gilbert had done was to capture polarity itself in the body of the terrella. What was true of the Earth in his private laboratory was conversely no less true of the actual Earth, and indeed all other similar heavenly spheres having magnetic poles. Here then was not only a philosophy and a theory, but an experimental body with powers of attraction

A depiction of 'versorium' or magnetic needle used by Gilbert with the lodestone to model the attraction of the compass to the Earth's poles.

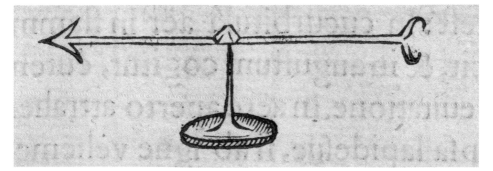

that promised to overturn the lazy half-truths that had been taught to him as gospel at Cambridge, and to reveal in entirely new ways the living relationships between the Earth's inner workings and the heavens.[6]

Gilbert, clearly a highly gifted experimentalist and observer, was also a philosopher exploring the metaphysics of life and the soul. Nothing less would do, he believed, if he were to establish the formal philosophical foundations for poles and polarity. Had Gilbert focused on solving the problem of the Earth's polarity, without speculating about the nature of polarity itself, he might be a household name today. Some of his most important ideas, however, would not stand the test of time. His fundamental insights would be challenged, reworked and developed by Isaac Newton and others, leaving Gilbert's legacy obscure. Gilbert did,

Gilbert's diagram from *De Magnete* (1600) shows how a ship's magnetized needle at different points on a terrella's surface (points C, E) will point towards the terrestrial North Pole (point A), unless deflected towards land.

Gilbert, likening the Earth to a very large terrella, showed pictorially how the principles of the terrella could then be applied to understand practical navigation, from *De Magnete* (1600).

however, understand something crucial about polarity that is easily overlooked. By finding answers to the problem of what poles really are, he was making poles more peculiar and remarkable, curious and inexplicable, instead of closing the subject with iron-cast conclusions. His new knowledge might appear to have contained the mysteries of the Earth in a laboratory model, but in fact his work was raising new questions that were exploding the old consensus of cosmography. Where in Ptolemy's cosmography the Earth was fixed amid the perfect harmonious movements of heavenly bodies, for natural philosophers this merely begged the question of understanding how exactly one substance exerted influence over another. Rival philosophies and explanations clashed. Gilbert's answer was that the Earth was suddenly exerting power over other bodies at a distance: a small distance in the case of his personal terrella, and far off as with the orbiting celestial bodies. In his magnetical philosophy the poles of rotating spherical bodies were lively, endowed with animate spirit or soul, and unrecognized and unaccounted for by the traditional laws of astronomy.[7]

This chapter tells the story of how the poles, by coming to life and breaking free from the old cosmography of the sixteenth century, multiplied in number, becoming more elusive and difficult to pin down. In this sense, the 'North Pole' in Gilbert's philosophy turned out not to be so singular: there were as many north poles as there were terrellas, not just the definitive points at the Earth's geographical pole or the celestial pole far above it. Polarity, freed from the early model of cosmography, would celebrate, intrigue and perplex philosophers in equal measure right up to the era of Romanticism in the early nineteenth century and beyond, as poles would continue to be associated with invisible and animated powers.[8]

One way to make sense of the rise of lively poles is to return to the idea in the last chapter about the importance of the North Pole in the design of certain practical instruments like Apian's speculum. This required observers to align their own bodies with the body of the instrument and that of the celestial axis. In that way the North Pole held together and coordinated the movement of the Sun, stars and other celestial bodies. A crucial feature of cosmographical instruments is that they were practical mathematical tools. Their makers and users cared little about why the instruments worked; whether their construction represented philosophical truths about the universe hardly mattered. It was simply enough that they worked, leaving the abstruse philosophical questions to theologians and natural philosophers. This is why specialists in the history of instruments tell us that the Copernican Revolution of the mid-sixteenth century made little impact on makers of practical surveying and navigational instruments. Philosophical and experimental instruments were a different matter, however, because they were being used to pose a different kind of question. Thus it was natural philosophers, not practical mathematicians, who asked the more searching questions, 'What is a pole and what makes a pole active or alive?' The terrella was conceived as a philosophical instrument to probe precisely these questions.

## William Gilbert, physician and experimentalist

Why William Gilbert became enamoured with magnetism can be traced back in part to his dissatisfaction with the curriculum at Cambridge University, where he had become Bursar of St John's College. Throughout his time there he had deplored and been highly dismissive of the dogmatic and uncritical truths taught to him and fellow students in the Aristotelian tradition. His was a complex personality: an iconoclast by inclination, but also a follower of the tradition of Renaissance Neoplatonists subscribing to the idealism of a perfectible human soul, and a passionate observer and interpreter of experimental evidence.[9] His was therefore no simple linear story of influences and discoveries. Posterity has cast him in different lights according to which aspects of his temperament and beliefs have been emphasized. Christopher Wren (1632–1723), for example, celebrated Gilbert as the founder of empiricism. He has also been called a 'semi-Copernican', because he subscribed to the view that the Earth rotates about its terrestrial axis – but without the full Copernican commitment to having planets orbiting the Sun.[10] What is known beyond doubt about Gilbert's bold and extraordinary study *De Magnete* (1600) was that it was the culmination of studies of magnetism he had been undertaking over thirty years.

*De Magnete* set out a comprehensive account of magnetic instruments, experiments and principles governing the behaviour of magnetic bodies ranging from needles and compasses to planets and the Sun. In his model, the Earth possessed a dipole, a north and south pole of opposite polarity, coinciding with the Earth's two geographical poles. Experiments showed how north and south poles attracted other poles of opposite polarity, and repelled poles with the same polarity. Though the celestial pole was believed by many mariners to be the source directing their compasses in the Ptolemaic model, Gilbert's poles put paid to the notion that any poles were passive, nor were they simply mysterious materials with hidden powers. Influenced by the Neoplatonists, Gilbert took the Earth's magnetic material to be 'ensouled', meaning it possessed a soul or life force that was active throughout the material. In *De Magnete* he explained that this magnetical soul was nothing less than the hidden cause of the Earth's rotation around its own polar axis. This principle of a soul causing rotation applied to some planets as well as the Earth.[11]

As strange as the idea of magnetic ensoulment seems to modern-day readers, stranger still was the fact that an experimentalist would turn to the metaphysical idea of ensouled poles to explain so much about the motions of bodies like the Earth. This suggested that the answer to the fundamental question – what are poles? – lay in the union of observation and philosophy. One has to keep in mind that two of the most important theories taught in school physics today simply did not exist in the sixteenth century: Newton's study of gravitation, which was published in the late seventeenth century, and the concept of a magnetic field, which only came about through the work of Carl Friedrich Gauss and James Clark Maxwell in the nineteenth century. As a result the poles took on enormous significance at the start of the seventeenth century because experiments showed them to be the most prominent and measurable of magnetic effects – and resistant to explanations. Thus Neoplatonists like Gilbert were going a step further when arguing that the poles contained a powerful soul-like substance that exerted a mutual power of attraction and repulsion with the souls of distant bodies.[12]

The pole was the focus of observation in Gilbert's conceptual framework. When he proposed that magnetism itself could exert reciprocal or mutual attraction over another magnetized body, whether a handheld piece of lodestone or a planet (at that time the Moon was considered a planet), the attraction was emanating from the poles. Of central importance to him was the idea that the path followed by distant bodies could be governed by mutual magnetic attraction acting at a distance with no intervening material, just empty space. Where philosophers inspired by Aristotle had viewed the universe as a set of solid spherical shells carrying and thereby causing the stars and planets to rotate around the celestial axis, Gilbert was one of a growing number of natural philosophers who believed Aristotle had been fundamentally wrong about cosmography, and that generations of astronomers and other supposedly learned followers had perpetuated his error-prone assumptions without bothering to examine them properly. This was a radical position to take because Aristotle's influence on the sciences and medicine had been enormous, if not all-pervasive.[13]

Gilbert was a radical and strident anti-Aristotelian in his practice of both medicine and natural philosophy. He believed that Aristotle's learning had become corrupted by the influence of his patron, Alexander the Great, who, as may be recalled, modelled himself on the Apollonian figure of the Hellenic god Herakles.[14] The Neoplatonists of the Renaissance believed that a true knowledge of the older philosophical traditions could be rescued thanks to the later reinterpretations of Plato, which essentially corrected the corrupted Aristotelian readings. Whereas Greek philosophers in the archaic age had been willing to grant active powers to the sub-lunar world, including the Earth itself, this was overturned by Aristotle, who placed the Earth firmly outside of the realm of divine virtue in the heavens. A Neoplatonist like Gilbert, restoring the possibility of divine virtue in the inhabited world of human affairs, could recognize signs of such power in the active poles of terrellas.[15]

Poles could do something particularly remarkable for a philosopher like Gilbert. Their power of attraction worked across or through the space between magnetic bodies, even over great (astronomical) distances. They exerted mutual influence without a material medium like an ether linking the bodies. Space in Gilbert's philosophy was a vacuum presenting no barrier to magnetic attraction, which had no place in Aristotelian thinking. It was also at odds with the mechanical philosophy of Gilbert's contemporary, René Descartes, who argued that spaces were filled with tiny corpuscles, so that forces acting at a distance could be reduced to a vast number of interactions over tiny distances. That magnetic poles could hold sway over iron needles anywhere on the Earth's surface, or across empty space to the celestial realm beyond, challenged the core tenets of philosophical rivals. Poles mattered!

The Moon, the nearest celestial body, was similarly animated by soulful material. Its path was largely determined by the mutual attraction (which eventually influenced Newton's conception of gravity) coming from the soul of the Earth. Gilbert realized that the Moon's soul pulls on the Earth's oceans, causing tides, which would in time be demonstrated by the Astronomer Royal Edmond Halley (1656–1742). Gilbert, ever the serious observer, drew what is believed to be the first map of the Moon based on lunar observations using the best instrument available to him – his own naked eyes! On Gilbert's lunar map, the North Pole is situated on a northern Arctic island, an *insula borealis*![16]

Gilbert's magnetized terrestrial poles were defined as points on the Earth's surface where the intensity of magnetism was greatest, but the irony was that the material comprising these poles was commonplace and dispersed. The magnetic substances, lodestone and iron, could be found in many countries, the most powerful lodestone being said to originate in China. Gilbert posited 'as many kinds of lodestones as there are different regions of soil in the whole world. For in every climate, province, and

The Moon's polar region with its *insula borealis* was first sketched by William Gilbert using his naked eye and was posthumously published in *De Mundo nostro* (1651).

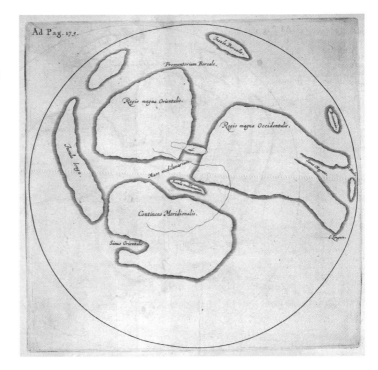

kind of land, lodestone is to be found or else lies unknown, deep, or inaccessible'.[17] Wherever there is soil, there is polarity.

Gilbert was a forensic reader with an excellent library and network of informants. He extracted whatever insights about magnetism he could from a field he reckoned to be dogged by error-strewn writing, uncritical, often plagiarized and all too often false. One plagiarized tract, however, proved extremely valuable. Published in 1572 as a posthumous work by one Jean Taisnier (1508–1562), it was later revealed to be a straightforward transcription of a very subtle and informed medieval study of geo-magnetism by a mid-thirteenth-century French military engineer named Pierre de Maricourt. He too had possessed a rare practical interest in carrying out his own magnetic experiments, which he had written up in 1271 under the Latinized nom de plume of Peregrinus. Maricourt's manuscript, or at least a copy of it, was given to his friend Roger Bacon at Oxford, where several copies were made.

Gilbert recognized in Maricourt a serious thinker and interlocutor, separated though they were by three centuries. Maricourt's treatise revealed an experimental approach that Gilbert would adopt, with a twist, to demonstrate his belief in an animated Earth. Maricourt had recognized that magnetic materials have poles: it is fair to credit him with the discovery of magnetic polarity.[18] He reasoned in the Aristotelian way that magnetism flowed from heavenly or celestial poles down into the magnetized material on Earth. Renaissance Neoplatonists would locate the source of magnetism in the nearest material body: the Pole Star. Gilbert expanded on this idea of a heavenly material magnetic pole, pulling it down to make it an earthly object. With Gilbert being a Copernican (or a version of it), instead of seeing the universe as moving like a machine, the rotating Earth needed its own set of ensouled poles and the strange powers that went with them.[19]

## Polar instruments and experiments

In order for a new and controversial polar theory of ensoulment to be compelling, it needed solid supporting evidence and Gilbert, sensitive to this, devoted a good deal of *De Magnete* to experimental demonstrations of his theory. Instrument-making in the sixteenth century was full of new innovations. Novel polar magnetic instruments were also to play a key role in Gilbert's construction of models and experiments by which he might envision a universe with a soulful and active terrestrial Earth. His crucial instrument, the spherical lodestone, came straight from the workshop of Maricourt, who likened it by analogy to a heavenly body on account of its perfect roundness and its polarity. In rechristening it a terrella, he was making explicit its status as a small or miniature Earth and not just an analogy.[20]

Did it matter that the terrella was spherical, unlike other magnetic shapes such as bar magnets? To be a miniature Earth, it of course needed the same perfect form. More than that, Gilbert needed to understand and demonstrate a central claim in his theory: that the poles are the cause of the Earth rotating.

Turning to Maricourt's work, Gilbert restated his argument that the magnetic soul in a spherical lodestone causes it to rotate slowly about its axis like the Copernican Earth. Gilbert took pains to demonstrate this through experiments, even though this proved to be a very difficult feat that others, as it would turn out, found impossible to replicate. However with the help of a very sensitive versorium as a detector, when brought near to a terrella the needle would move in a circular motion, demonstrating how magnetic attraction moved in curved or circular trajectories, not straight lines. The implication was that the curved path followed by the needle revealed the rotational character of the lodestone's magnetism acting on the needle, and therefore reciprocally that of the needle on the lodestone – the miniature Earth.

Gilbert was not alone in looking to pole-seeking magnetic instruments to advance an experimental understanding of the Earth. His contemporary Robert Norman (*fl.* 1581), navigator,

Johannes Stradanus' depiction of Flavio Amalfitano's study (*c.* 1580–90) illustrates how the mystery of polarity concealed in the lodestone holds the key to understanding the compass, navigation and the globe.

2.                     LAPIS POLARIS, MAGNES.
*Lapis recluſit iſte Flauio abditum     'Poli ſuum hunc amorem, at ipse nauita.*

A declination instrument, later called a dip instrument, enabled navigators to determine their latitude by measuring the downwards angle of the pole (inside the Earth) from the horizon. From *De Magnete* (1600).

instrument maker and Neoplatonist, developed a highly original and practical instrument called the dip needle. Unlike the compass needle, which rotates more or less horizontally with a card to indicate azimuthal direction, the dip needle was mounted so as to enable it to swing freely in all directions including the vertical plane. This opened up the possibility of locating the poles inside the Earth instead of on its surface. The dip needle could measure the angle between the horizon and an interior magnetic pole using a scale adapted from an astrolabe. A travelling observer, whether a mariner or surveyor, equipped with a dip needle could similarly follow the magnetic needle when it pointed beneath the horizon dipping into the Earth. By voyaging to find the place where the needle points straight down into the ground, the dip needle would mark the exact position on the Earth's surface above the magnetic pole, however close or far from the geographical pole that might be. This in fact is what the dip needle was able to do – nearly 250 years later when James Clark Ross (1800–1862) stood over the appropriate location in the Canadian Arctic in June 1831.[21]

Gilbert reasoned that the poles on the Earth's surface were the focus of the Earth's ensouled magnetic power, but that the material was diffused throughout the Earth according to the deposits of magnetic lodestone or iron. If this were true, it followed that the magnetic poles were not geographically unique sources at all, just locations where the attraction was most intensive. This kind of pole was not then privileged like the uniquely sublime celestial Ptolemaic pole; nor was it the embodiment of a soul hidden locally in one place at the top of the world. To the contrary, the magnetic pole was the cumulative effect of ensouled material that was widely distributed over the world's continental surface and in its interior (but not the oceans) according to laws unknown following no obvious pattern.

The failure of the compass needle to point to the geographical North Pole at 90°N implied either that compass needles were misbehaving or that the magnetic poles lay somewhere other than at the geographical poles. Gilbert reasoned that degraded magnetic material in the Earth's continents was interfering with the compass needles. The two measures of this distortion, magnetic variation (horizontal) and dip (vertical), could therefore in principle be tamed and even turned to advantage if they could be measured and mapped. An appreciation of the complexity of these distortions was growing fast. Within a generation of Gilbert's death, secular variation – the changing deflection of the Earth's magnetic field over time – had been established by a series of experiments in London between 1622 and 1633. Thus careful experiments on terrestrial magnetism, while revealing remarkable new truths, had also raised profound complexities that stood in the way of a comprehensive and coherent account of poles and polarity.

## Mapping the poles

Magnetic variation and dip defined the principal problems for research into terrestrial magnetism for two centuries after the publication of *De Magnete*. If the magnetic pole wasn't at the geographical North Pole, then it could only be discovered if navigators, instrument makers and philosophers joined forces. Only then would they understand which instruments to take, how to use them and what to look for. Serious navigators knew that variation (the angle between the geographical and magnetic poles) changed in unspecified ways over long distances, casting doubt on their trust in their compasses. Part of the solution was for navigational instruments to be made simple and robust enough to be reliable on the high seas. But there was another more nagging question, the cause of magnetic variation itself. The realization that this variation between true (geographical pole) and magnetic north could occur over even the course of a day (secular variation) resulted in a compound problem that had instrument makers and natural philosophers tearing their hair out in frustration. Instrument makers needed to assure their customers that their instruments worked, while natural philosophers needed to know why instruments didn't work.

The solution was to team up. At Gresham College, a precursor to the Royal Society, practical mathematicians, instrument makers and philosophers could meet and work together. Their brand of collaboration was to travel across workshops, houses of experiment and ships' decks in far-flung oceans. Emerging out of this was a new model for organizing research expeditions that would define polar voyages of discovery for centuries to come.

When Captain Thomas James (1593–1635) set sail in the *Henrietta Maria* in 1631 to discover a Northwest Passage, his privately bankrolled expedition was fitted out with a remarkable battery of the most advanced navigation instruments, among which magnetic instruments figured prominently. James was supplied with a dazzling array of instruments of many kinds and purposes. The quadrants, Gunter's cross-staff, and 6- and 7-foot cross-staves were chosen for navigation. Others, like the 'Chestfull

Joseph Moxon in his *A Brief Discourse of a Passage by the North-Pole* (1674) presented evidence gathered to persuade readers that a passage by the North Pole was possible because Dutch whalers had already purportedly reached and sailed beyond it.

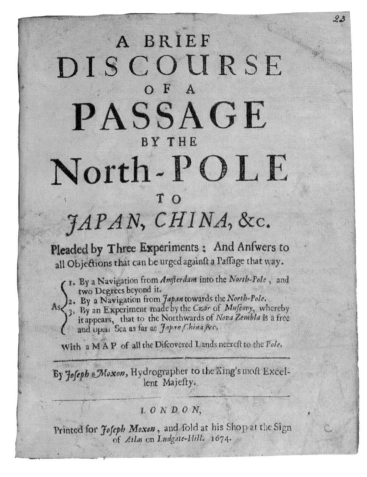

A BRIEF
DISCOURSE
OF A
PASSAGE
BY THE
North-POLE
TO
*JAPAN, CHINA,* &c.

**Pleaded by Three Experiments :** And Anſwers to
all Objections that can be urged againſt a Paſſage that way.

As
1. By a Navigation from *Amſterdam* into the *North-Pole* , and two Degrees beyond it.
2. By a Navigation from *Japan* towards the *North-Pole.*
3. By an Experiment made by the Czar of *Muſcovy,* whereby it appears, that to the Northwards of *Nova Zembla* is a free and open Sea as far as *Japan China &c.*

With a M A P of all the Diſcovered Lands neereſt to the *Pole.*

By *Joſeph Moxon,* Hydrographer to the King's moſt Excellent Majeſty.

*L O N D O N,*
Printed for *Joſeph Moxon* , and ſold at his Shop at the Sign of *Atlas* on *Ludgate-Hill.* 1674.

of the best and choisest Mathematicall books, [the best] that could be got for money in England', were dedicated to surveying and astronomy.[22] Of lasting importance to this and subsequent polar expeditions were the *Henrietta Maria*'s magnetic instruments: 'six meridian compasses, ingeniously made; besides some doozens of others, more common. Foure needles in square boxes, of six inches Diameter . . . moreover foure speciall Needles . . . of six inches diameter.' These were no mere needles; they were exceptional. They had been magnetized or touched 'with the best Loade-stone in England'. In case they needed retouching during the expedition, a special loade-stone was provided, with the precaution that its north and south 'Poles . . . [were] marked, for feare of mistaking'.[23]

The Arctic was a harsh, beguiling and unforgiving testing ground. Royal Society–Admiralty expeditions followed the Thomas James expedition model to the frontiers of empire, bringing back much-prized readings of magnetic variations.[24] Discerning the difference between a bad instrument and a bad observer was difficult and often insoluble. Aggravated by having to hold instruments in conditions of extreme cold, tracking magnetic variation was inherently difficult, as it increased steeply as a ship drew nearer to the magnetic pole. Compass needles were reported to behave in strange and worrying ways, circling one way one moment and back the next. Variation seemed prone to changing unpredictably, producing confusion and consternation rather than the hoped-for discoveries. The reputations of observers and instrument makers were also being tested, and in extremis to the point of breaking. While Gilbert could voyage to the polar regions in the comfort of his workshop, the truth was that the biting cold of the polar Arctic would long remain a notoriously difficult environment in which to handle instruments so that they might work reliably. The rapid shifts in seasonal light played tricks on the landscape, fooling one's perceptions. Magnetism, largely invisible yet ever present, jealously fended off attempts by gentlemen of science to reveal the secrets of its nature beneath the surface of the tundra and polar sea ice. These forces – light, cold and magnetism – produced an unforgiving and fickle power exuded by the Earth, flattering observers only to deceive, filling many journals with anguish and disorientation.

Joseph Moxon (1627–1691), hydrographer to Charles II, described in a pamphlet on polar navigation how, to deal with polar disorientation, ships 'must needs lose themselves in the *North-Pole*'. The disorientation experienced in high latitudes was logical and paradoxical because the compass needle 'pointing always North . . . must indifferently respect all points of the horizon alike'.[25] There, standing over the magnetic pole, the dip needle could point down – but how far down was the pole and what was the nature of its power? The dip needle, when aligned

Joseph Moxon published a polar stereographical chart showing the North Pole in the centre, thereby making the idea of a polar passage more concrete and plausible (1674).

with the Earth's pole, was the hallmark of standing on the pole, but it also signalled a loss of direction; on that spot, the compass needle could only point south or rotate to and fro in pattern-less uncertainty. This was no different from standing at the geographical pole, every direction being south. The principle governing both is that a system of spatial orientation requires distance between the observer and the landmark; when that distance disappears, so does its use for orientation.

What if the pole-seeking needles of dip circles were being drawn towards not one but two or possibly more interior poles?

The young astronomer Edmond Halley presented his analysis, or at least a conjecture, to the Royal Society in 1683 that the measurements of variation reported by navigators, when taken together, indicated the presence of not one but two pairs of poles. Halley gave the four poles place names. He used the names of the continents closest to where the poles were dominant: the European (Spitsbergen) and American (Bering Strait) North Poles, and the American South (Pacific Ocean) and Asian South (in the Southern Ocean south of modern-day Sulawesi in Indonesia) Poles. In the polar and temperate zones, each pole exhibited dominance in its own vicinity and region. At certain locations on the Earth's surface, the attraction of the poles was finally balanced, giving zero variation. The line connecting these points was called the 'agonic' line of zero variation. Halley speculated that the behaviour of magnetized needles was governed by the 'counterpoise of the forces of the two magnetic poles of the same nature'. In the torrid equatorial zone all four poles were deemed to hold sway in some measure or proportion.[26]

Through the work of Isaac Newton (1643–1727), concepts of force and gravitation became much better understood in the late seventeenth century. Polar magnetism, however, remained stubbornly difficult to explain, even as Gilbert's idea of 'ensoulment' was dropped as mechanical explanations held sway. Halley nevertheless continued to argue the case that mapping the contours of magnetic variation was the best solution for developing a reliable method for finding longitude at sea, if only navigators could build up a sufficient stock of magnetic observations. Halley admitted to having doubts about the possibility of ever satisfactorily confirming the validity of the four-pole model without decades of further painstaking observations by navigators. By identifying the regions in which variation seemed to change most over time, Halley advanced a new and more complex hypothesis to explain the uneven shifting patterns of variation and dip: one pair of poles – European, American South – were fixed in the Earth's outer solid shell or cortex, while the other pair (American North, Asian South) moved over time through the

J. Faber's 1722 portrait after T. Murray of Edmond Halley, Astronomer Royal and Savilian Professor of Mathematics at Oxford University. Halley sought to harness the observation of past, present and future mariners to map what he believed to be the Earth's four interior magnetic poles, each pair situated on one of two concentric inner shells separated by an atmosphere.

Earth's fluid magnetic nucleus. As the temptation existed to marshal observations to fit theory, the putative location of the European pole also shifted from Russia to the polar sea north of England.[27]

Things were not getting any easier for any navigator unsure whether his compass needle was pointing to one pole, or being pulled by the forces of two or four poles at any one time. Halley, however, persisted in his belief that the movements of the magnetic poles could be observed and in the course of time decoded and then understood as a system. In the meantime, he published his first isogonic chart showing lines of equal variation across the

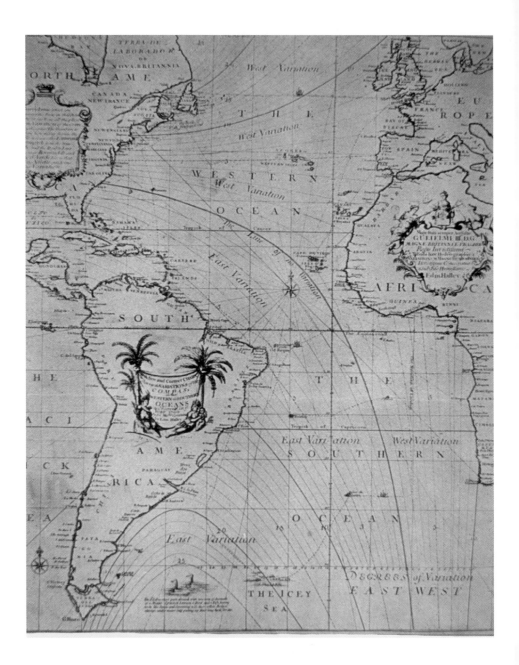

Edmond Halley developed his *New and Correct Chart shewing the Variations of the Compass* (1701) in the belief that magnetic variation could eventually become a reliable method of determining longitude at sea (though this would not work in those areas where the lines are horizontal).

globe. Where the lines closely followed north–south meridians, navigators sailing east–west would be crossing the lines and able to use Halley's chart to determine longitude. Mapping continually changing isogonic lines on the scale of the globe, while an inspiring idea, was too great a task to organize. Making matters worse, the latest observations coming to Halley suggested that the poles in the northern and southern hemispheres were not actually diametrically opposite each other. If north and south poles were not polar opposites, how were they connected?

Shrouded in uncertainty, unseen and shifting, the poles presented a very different picture than the one Gilbert had envisaged a century earlier. Novel instruments and experiments had helped unleash a philosophical world that was difficult to contain. Fluid magnetic lines operating inside the Earth spawned new kinds of mapping expeditions. Ships travelling over the Earth's surface followed instruments guided not by the law-like truths of fixed bar magnets or terrellas, but instead a perplexing dance of moving poles. One hundred years on, Gilbert's confidence in navigation by terrella looked to have been badly misplaced.

So much about the poles had changed over 150 years. The Werner and Apian generation of cosmographers, committed to Aristotelian and Ptolemaic principles, described a universe of six eternal and unchanging poles: two celestial, two ecliptic (poles of the Sun's orbit) and two geographical. With the arrival of Copernicus' heliocentric universe, each orbiting body possessed its own separate polar axis of rotation. Though the celestial poles remained of great importance in the design of instruments, models and astrology, they had lost their ordained role defining the universe's sole axis. The animism of Neoplatonists like Gilbert had inverted this order, identifying divine virtue in magnetized material itself. Every magnetized body, including the Earth, Moon and Sun, generated its own self-motion through its respective poles, bringing the pole count to at least twelve.

For Neoplatonists like Giordano Bruno (1548–1600), the possibility of new poles and new worlds went hand in hand. His argument that, in principle, an infinite number of possible worlds could exist, by extension also argued that poles, and therefore life itself, could exist ad infinitum. This radical philosophical pluralism cost him his life. Condemned to death by the Church for his heresy, he was burned at the stake. Gilbert's magnetical experiments had revealed that poles, far from marking out the universe's divine harmonies, resided in the material of any piece of lodestone, albeit lodestone with soul. On that reading, poles could be created indefinitely by taking a piece of lodestone from a mine and crushing it into pieces – not very exalted after all. That said, Gilbert, though neither the first to make a lodestone nor to touch (or magnetize) an iron needle, had done something very special. By demonstrating the means to create new poles, he had shown experimentally, or so he believed, that any magnetic spherical body would rotate on its axis.

The experiments of Gilbert and his contemporaries had also opened up to study the Earth's highly dynamic interior movements. Interior magnetism had pointed to the presence of a generative power within the Earth directed outwards through its poles. Like a gate through which this attractive power could travel out over and beyond the Earth, magnetism could attract distant bodies possessing similar substances. Hence the poles could project and connect bodies across enormous distances, linking the Earth's interior and exterior. The breaching of the Earth's surface at the poles, as much as they puzzled natural philosophers, would soon find fertile ground in the imagination of literary-minded philosophers, utopians and writers of many genres, a theme we shall return to in the next chapter.

# 4 Polar Voyaging

'Among the enterprises which yet remain unaccomplished, and of which the object is to complete our knowledge of the surface of the globe, it appears to me that there is none more desirable ...and few so easily practicable, as an attempt to reach the North Pole of the earth.'[1] So wrote Edward Parry, Britain's most successful polar explorer, in the run-up to his 1827 attempt on the geographical North Pole. Such unguarded optimism prior to an expedition of this kind is curious. After eight years exploring the intricate maze of dead ends in pursuit of a Northwest Passage, pursuing this new goal was perhaps refreshing and less complex. Parry was not alone among experts in believing that the odds were on his side. William Scoresby Jr, whaler and naturalist, and Britain's most experienced and authoritative Arctic navigator, had a decade earlier in 1815 ventured 'that a journey over a surface of ice from the north of Spitsbergen to the Pole' might meet with 'a probability of success'.[2] Fifty years earlier, Daines Barrington, Vice-President of the Royal Society and an armchair geographer, had judged the chances of a seaborne passage to the pole using a pair of warships as 'probable' and 'practicable'.[3] Seemingly it was just a matter of time before intrepid navigators would close in on the pole.

Such uniform expectations of success may have been more a case of advertising and hype dressed up as scientific reasoning to win supporters for this fantastic venture. In fact the optimism belied deep differences about the polar quest. The question of how to reach the North Pole turned in large part on the more

Charles Knight, *Daines Barrington*, 1795, engraving after Joseph Slater, 1770. Barrington was a wealthy gentleman antiquarian, naturalist and armchair geographer, and a powerful advocate of northern exploration.

uncertain question of climate and ocean conditions: would the pole turn out to be an island, open sea, a sheet of smooth ice, jagged ice or some combination? Nobody knew, but as an exercise in erudition and speculation, it was a laudable pastime for an extremely wealthy gentleman like Barrington. Geographers since the early days of the Royal Society and on into the late nineteenth century had argued that the polar ocean was ice-free. Learned men like Robert Boyle claimed that sea ice built up adjacent to land masses, but would not be at the geographical North Pole if it were far from land. In 1815 Scoresby, who had diligently studied Arctic natural history at first hand over many years, published an erudite study of the different kinds of ice, their formation and structure. He and his professors at the University of Edinburgh were in little doubt that all he had seen during his many seasons of high-latitude whaling pointed to very considerable amounts of ice surrounding the pole. Not everyone agreed. John Barrow, Parry's patron, favoured the open polar sea theory and ridiculed Scoresby's views, describing them as an 'idle and thoughtless project' and the 'frenzied speculation of a disordered fancy'.[4] In the end, Parry's expedition reached 82° 45' and no further thanks to the sea ice drifting nearly as quickly southwards as he could sledge northwards. Scoresby had been proved correct.[5]

As much as one might expect the conquest of the North Pole to have been settled once and for all, that never really happened.

This map of the polar sea with a passage between the surrounding continents was indebted to the speculative mapping of the French cartographer Philippe Buache (*Gentleman's Magazine*, 1760).

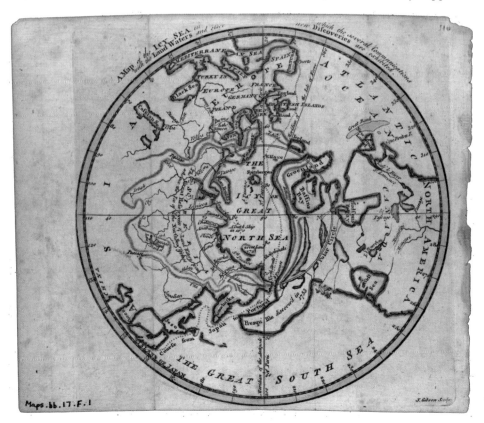

At no point in time did the world agree unequivocally that the North Pole had been discovered or definitively tamed. Instead, over the course of the next two centuries the goalposts would be moved, the challenge would be redefined, new technologies of travel would be introduced, aviators and submariners would take part, and the creative reinvention of what counts as exploration would continue unabated. In this way, polar expeditions often retained an idiosyncratic character. Being pole-bound would remain the exception throughout the history of Arctic voyaging, never becoming rooted enough to be a tradition, or the rule.[6]

The earliest North Pole voyages proposed by Robert Thorne (1527), Hugh Willoughby (1553), Willem Barentsz (1596) and Henry Hudson (1607) had been envisaged as possible trade routes to Cathay. The idea of a pole surrounded by open sea was a commercially attractive proposition, as it held open the prospect of a new trade route and potentially great wealth. The practical limits to navigation were eventually established by those who annually made the Arctic seas their summer home, the whalers who built their huts and set up cauldrons for boiling blubber on the shores of Spitsbergen at 80°N. In some years the wall of ice to the north of Spitsbergen opened up, receded or fractured, then the more intrepid whalers made forays further north; other seasons the ice hugged the shores and the whalers stood off. What lay north was a matter of local knowledge, hearsay and anecdote circulating among the community.[7]

From the voyage of Constantine Phipps (1773) onwards, attaining the North Pole became more a test of speed, skill and tactics, and less the pursuit of the early modern dream to rival the Spanish and Portuguese with a new route to Cathay. The sciences of natural philosophy and magnetic studies were part of the plan too, on the model of Thomas James's partnership between the Admiralty and Gresham College. Phipps's voyage had in practice morphed into a glorified 'steeplechase', conceived as single-season exploits, a dash out and back to be home before the onset of winter.[8] The spirit of empire and conquest was certainly present, but now the key parameters were speed, design, timing and tactics.

This phase of maritime discovery reached its imperialist zenith when a group of men of different ethnicities photographed themselves atop the world on a hastily constructed stage of snow in the vicinity of the pole. This is a short chapter in the story of the geographical imagination of the Earth: sporadic events spanning little more than 150 years of actual exploration. Seeking the geographical North Pole in the late nineteenth century was reimagined as a race and an expression of imperial desire. It could also be understood as the crowning achievement of an Enlightenment project of global exploration beginning with Captain Cook's circumnavigations, the ultimate survey of nature's geography at its harshest, and a symbolic acting out of territorial power. In truth there was nothing final about sailing to the North Pole. These voyages took place principally at moments in history when the protagonists – Britain, the United States, Norway and Russia – were in the imperial ascent or seeking to reassert political prowess. Riding the crest of a wave of nationalist ambition, the pole's symbolic value tended to rise and fall with national political and economic fortunes and anxieties.[9]

North Pole voyages, however esoteric or geographically removed from society, nevertheless mirrored the materials and organization of industrial societies that produced them. Exploration ships like Phipps's *Racehorse* and Parry's *Hecla* were bomb vessels, powerful warships adapted to pounding the ice instead of enemy vessels. The navigation instruments included the most advanced technologies that artisans and men of science could muster. Phipps had been entrusted by the Commissioners of the Board of Longitude to take Kendall's H2 chronometer all the way to the North Pole to see if it could keep time in such an extreme maritime environment.[10] Subsequent voyages would incorporate iron into their constructions of ships and boats. The leather-bound exploration narratives would be printed in the nineteenth century using advanced steam-powered presses to reach a new burgeoning mass market for reading. For all that polar voyages were often romanticized as a confrontation with pure nature, they self-consciously boasted the hallmarks of industrializing societies.

The model of sailing to the pole and back in a season had its roots in the voyages of the Enlightenment. Cook's French counterpart in circumnavigating the globe, the Comte de Bougainville, shared a proposal he had drawn up for a voyage to the North Pole that failed to win financial backing. Soon afterwards, in 1773, the British Admiralty commissioned Constantine Phipps to sail for the North Pole in the *Racehorse* and *Carcass*. Were he to reach the North Pole, his instructions required him to return to Britain without delay, and not to sail beyond the pole to the Pacific. Phipps's voyage narrative (1774), like those of his contemporary Captain Cook, was written collaboratively, not individually. The visual and textual narratives of their voyages provided rich raw materials that inspired the European Romanticism of early nineteenth-century writers including Mary Shelley, Samuel Coleridge and Robert Southey.[11]

The formidable barrier of ice impeding the passage of military war vessels like the *Racehorse* and *Carcass* (1773) became a recurrent theme of British North Pole approaches by sea. Illustration by John Cleveley the Younger, 1774.

## The North Pole prize

Any geographical test worth its name required a reward system with prizes and a set of standards. In 1776 Parliament introduced a huge inducement of £20,000 for the discovery of the Northwest Passage, the same size as the existing reward for discovering an accurate method of finding longitude at sea. A smaller but still very sizeable sum of £5,000 was allotted for sailing to within one degree of latitude of the North Pole. Crossing the 89th parallel was initially deemed close enough to the North Pole to claim the prize, but it also turned out to be too intimidating or far out of the way to tempt those to whom it was principally directed, the owners and captains of Arctic whaling ships. Taking up the challenge, they worried, would require such a detour from the whaling grounds as to lack any commercial logic or worse still to invalidate the insurance on their ships, It was a risk too far, not worth the trouble, and consequently the prize was an ineffective incentive.

When the Admiralty renewed its polar expeditions in 1818, a new set of polar prizes was offered, rolled into the Longitude Act, which incorporated a 'latitude act' in all but name – like a Northwest Passage defined by its northern rather than western

The land closest to the North Pole was the island of Spitsbergen at approximately 80°N, home to European whaling bases since the 16th century. Frequently sketched and painted, it was often used to represent the North Pole, as here by P. Auvergne, 1774.

boundaries. Why the North Pole should deserve a reward separate from that for the Northwest Passage is an important distinction that reveals a different logic of exploration.

The wording of the 1776 parliamentary Arctic discovery act provides a clue. The term 'approach' is used to describe a navigation to the North Pole, whereas the term 'discovery' is reserved for the completion of the Northwest Passage. Similarly, claimants for the North Pole prize are called 'approachers', those for the Northwest Passage prize 'discoverers'.[12] Rewarding someone for approaching rather than discovering a sea route to a place was less remunerative; ostensibly a lesser kind of achievement, it pointed to a different kind of navigation. A northern passage between the Atlantic and Pacific, if well documented by credible astronomical observations and log books, could be a definitive discovery, the holy grail of geographical enquiry revealed. But the North Pole's location was by definition perfectly known, and was yet an unlikely destination. So what aspect of a pole-bound voyage would count as discovery? Even though a transpolar route sailing down into the Pacific Ocean could be construed as an unambiguous discovery, how would a claimant for the North Pole prize be able to show that the precise location, an infinitesimal point, had actually been reached? How near was near enough, or would an approach always remain just that, never quite arriving, never actually being there?

This paradox of approaching was very emblematic of a society in which precision measurement was becoming more and more important. The North Pole was the cartographic origin point of longitude, the place from which its radiating lines emanated, and therefore the point at which longitude could not be measured. The search for longitude had become a popular source of satire and scepticism among eighteenth-century literary figures and philosophers like Jonathan Swift. It amounted both to a cultural argument against polar discovery and a source of fantasy, a gesture towards what lay beyond the pole. Of course the paradox of the pole really did matter to cartographers, who had to make practical decisions about projections and distortions. Even as simple a problem as how mariners following compass bearings could follow their ship's course on a globe demanded an understanding of the polar paradox. Steering a steady northerly course (such as northeast at 45) would, if transposed onto a globe, trace out a spiral circling the pole. This 'rhumb line' at sea would always approach the pole without ever actually arriving. In fact following any bearing would, if unobstructed, in theory eventually lead one to the North or South Pole. In that sense, the one place that all roads lead to (excepting east–west courses that follow a single line of latitude) was the place that no one seemed able to reach.

Daines Barrington recognized that paradoxes could stand in the way of North Pole expeditions, either because people might believe them or simply enjoy them more than having them exposed. Thus he took to print to argue against such popular objections. Even steering due north, 'approaching to a passage under the Pole', explained Barrington impatiently, was 'treated as paradoxical by many' because they believed it would 'destroy the use of the compass'.[13] By then it had long been known that compasses didn't point to the geographical pole. Even if scientific enquiry since Copernicus and Newton had shown that going out and looking – geographical empiricism – was hardly guaranteed to produce unambiguous discoveries, it was also acknowledged that maritime exploration was an essential experimental vehicle for the sciences. This was a good argument to put forward to help justify polar expeditions as serious knowledge-producing endeavours of lasting benefit.

Barrington's use of the term 'possibility' in the title of his treatise tells us something about the nature of North Pole proposals and expeditions. Scientific authorities like Robert Boyle had argued that sea ice could form in bays and the mouths of rivers, but not out at sea far from coastlines. Barrington argued that a passage was probable in the sense of being very likely ice-free and therefore unobstructed. The practical challenge was then to find a way to sail past the barrier of ice ringing the Arctic Ocean along the coastlines of Russia, Spitsbergen and Greenland, to reach what theory suggested would be an open polar sea.[14]

Barrington, an antiquarian, was obsessive in collecting information about previous northern voyages. He excavated plenty of material suggesting that whalers had already been near the North Pole, or at least in very high latitudes: no fewer than six vessels were said to have reached 86°, three to 88°, two to 89° and one to have reached 89½°. How reliable the sources were was hard to say, as in most cases proof was lacking: logbooks and astronomical observations were deemed to be the difference between optimism and demonstration. A generation later, William Scoresby Jr followed suit in his authoritative *Account of the Arctic Regions* (1820), arguing that none of the 'farthest north' achievements cited by Barrington could be sufficiently verified by observation and logbook. Mainly whalers, these men didn't stand accused of any dishonesty; it is that their stories were just that, unreliable and prone to exaggeration. In this way, the judges for discovery prizes, defining the rules of the prize money in terms of their own contemporary standards of precision measurement, were able to wipe clean the slate of past navigators who had already claimed to have approached the pole – and indeed may have.

Barrington's second-hand reports clearly had some impact because the Commissioners of Longitude were persuaded to set an exceedingly demanding benchmark of 89°N for the prize of approaching the pole. Taking into account that the Board of Longitude's learned astronomers were well equipped to scrutinize and pounce on the slightest fault of anyone petitioning them with solutions to the longitude problem, in the expectation of detecting fabrications and incompetence, alongside those of the self-deluded, the North Pole was truly out of reach of most navigators. The result was that the North Pole prize did little to persuade potential approachers to risk everything on reaching the 89th parallel.

The government eventually realized that it needed to change the rules if it wanted the polar quest to be taken seriously. Taking advice from Scoresby and Joseph Banks, President of the Royal Society, the government reformed the Act in 1818 to allow for proportionate rewards. The idea then was to break down the

North Pole and Northwest Passage into stages to replace the all-or-nothing status quo: for example, the North Pole reward was set to begin at 84°N, offering £1,000 for each additional degree of latitude, progressing to £5,000 for ultimate success beyond 89°N.[15] Banks and Barrow weighed up the pros and cons of a Northwest Passage search versus a North Pole approach. They hedged their bets and opted for both, outfitting four ships, two for each goal. After the North Pole ships, pummelled by the ice, barely managed to limp back to Britain, polar approaching gave way to exploration for a Northwest Passage. Darting through polynyas and cutting tracks through thick sea ice was arduous, sometimes perilous and at other times tedious, requiring methodical guesswork amid a mass of islands with mostly ice-filled coastal waters. These expeditions into the heart of the Inuit homelands rarely ventured more than a few degrees beyond the Arctic Circle. They were almost always hugging a northern shoreline, never traversing the ice barrier into open sea.

During this period (1818–30) the mood for scientific reform turned sharply against the kind of state-sponsored science and patronage inspired by James, Phipps and Cook, where the government could allocate substantial sums of money to expensive expeditions in the name of science. The credibility of polar exploration became associated with a narrative about the 'decline of science' amid bitter accusations of nepotism.[16] After Parry's 1827 failure to reach even 83°N the Board of Longitude and the Longitude Act, along with the polar prize money, were abolished at a stroke. It would be half a century before the Admiralty would try again with the Nares expedition.

These were really a sequence of complex cosmographical bets or gambles. It was part of the human condition or predicament that we had throughout history been confined to the surface of the globe. Even with scientific experimentation and observation, it was difficult to escape this surface life to discover great truths behind constellations and geological forces beyond our purview. The Enlightenment philosopher Immanuel Kant described this as a search for *orientation*. To know where we are, he argued, was as much a question of knowing who we are. Exploration of the

J. Thomson, *Edward Parry*, after painting by S. Drummond, 1820. Parry was by the time of his North Pole expedition in 1827 Britain's most famous explorer after the immortal Cook. Having commanded three Northwest Passage expeditions, he had acquired enormous experience in Arctic navigation as well as a very profound if flawed understanding of the culture of Inuit peoples.

globe would always require improvising, a messy business, never a perfect science. The notion of an absolute framework, a truly objective perspective looking down on the world, was simply utopian and illusory. The desire to stand at the North Pole captured the desire to master our geographical existence, to find a solution to our painful predicament of being earthbound.[17]

If the North Pole could be navigated using the understanding of magnetic instruments, as Gilbert had argued, linking the heavens and the Earth's deep inner soul, then it was also an emphatic recognition of the severe limitations of the realities of human mobility, unable to move very much above or beneath the Earth's surface. For polar planners and explorers, approaching the Earth's poles demanded that they push mobility to its limits. Planning,

Mark Beaufoy's 'Map of the Countries Around the North Pole' (1818). Early 19th-century geographers and hydrographers like Beaufoy, Barrow and Hurd renewed the cosmographical and imperial vision of stereographical polar maps, again highlighting America, Europe and Asia, like Apian three centuries earlier.

organization and discipline, tenacious endurance coupled with new technologies – all these were key ingredients. The most adroit explorers also appreciated that powers of concentration in observation, attention and listening to the changing faces of Arctic nature could make the difference between success and failure. Being willing to adapt strategies to conditions on the hoof was essential. Only then might one come to know whether the theory of a warm, ice-free polar sea would turn out to be true or discarded on the heap of ill-founded armchair conjectures. This was the creed of geographical empiricism: search, observe, record. The truth would be revealed by witnessing at first hand.

Planning pole-bound expeditions meant giving due thought to technological design and strategy. Polar expeditions frequently

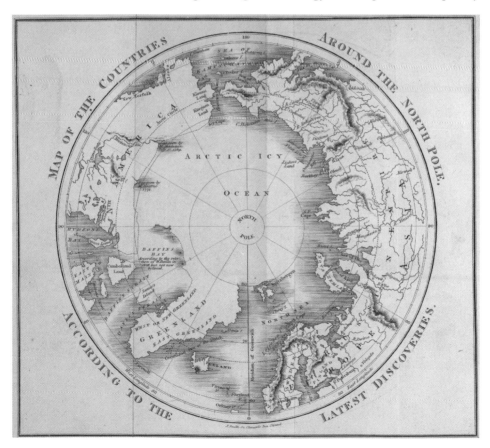

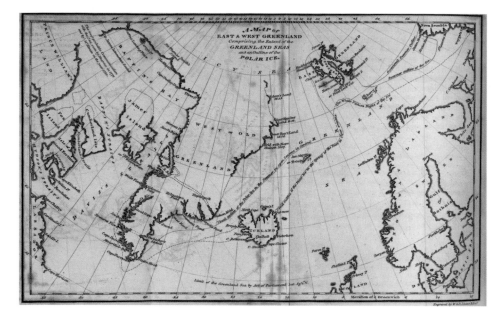

incorporated distinctive design features such as steam engines or ice-strengthened hulls, sometimes at the same time as drawing attention away from leaky ships, malfunctioning instruments and deserters among the crew. In this respect Parry's seventy-day dash to the North Pole was a landmark in expedition design. The four Northwest Passage expeditions in which he had played leading roles had given him enormous personal experience of navigation in Arctic waters. Though he confronted the Arctic ice with two heavy ships of war and correspondingly large crews, he had learned a great deal from the Inuit. His expeditions had gained practice making reconnaissance excursions using small crews travelling by sledge and knew at first hand that they were nimble and flexible in crossing different terrains of snow and ice. Parry had also grown tired of maintaining discipline among his officers and seamen and longed to travel with a smaller hand-picked crew. Discouraged by not having found a northwest passage, the sledge journey to the North Pole offered a welcome diversion or alterative. A warship would be used to get to the Arctic and to deliver the sledge party to its launch site on the north coast of Spitsbergen, with the necessary supplies.

William Scoresby Jr, whaler and polar naturalist par excellence, published this pioneering map in 1815 showing the seasonally changing contour of the polar ice edge, forming as far south as the southern tip of Greenland (58°N) and as far north as the Seven Islands of Svalbard (81°N).

It would also provide a safe platform for scientific experiments while he was away. Those considerations aside, the plan was to be small and light.

Parry's sledge-led journey was based on a plan proposed by William Scoresby Jr more than a decade earlier. Parry, like Scoresby, had spent enough time in the Arctic to scoff at the theory of an open polar sea or, at the very least, to have serious reservations about it. Scoresby had therefore proposed a sledge design pulled by dogs or reindeer capable of making good speed over the polar sea ice. A key condition was to depart early enough in the spring season before the ice pack broke up, to maximize the continuity of the surface over the ice. Parry had also experienced a wide range of ice conditions, including the pressure ridges formed by ice fields or floes being forced together, as well as unexpected bodies of open water. Thus it made sense to follow Scoresby in proposing a versatile solution, a 'light-weight

This illustration depicting Scoresby's crew desperately rescuing the ice-damaged keel of his ship, the *Esk*, shows how precarious polar navigation could be, even for the most experienced of whaling captains.

part abundant, with seals, walrus, fish and whales feeding on the krill and plankton that thrive in the mixing of warm and cold sea currents. The inland caribou herds were equally an indispensable source of winter clothing and meat hunted by Inuit during the summer season when the sea ice was retreating. Over the course of about eighteen months of intensive socializing with the Inuit from the Igloolik region, Parry and his second in command, George Francis Lyon (1795–1832), had become intensely interested in the culture of the Inuit, their language, mobility, sources of sustenance and belief systems. The routes to the North Pole, by contrast, seemed to be another world altogether, leaving behind the habitable world of the Inuit with few signposts of the trail ahead, bereft of the stories indicating how to find and travel safe trails. For Parry, going for the pole meant stepping out of the social world of the Arctic that he knew best and heading off north into the uninhabited unknown, where people and sea mammals were scarce. Even the islands on the north side of the Northwest Passage were rarely visited by Inuit. At higher latitudes, the North

North Pole expeditions almost always proved more arduous in practice than made out during planning or fundraising. 'Travelling among Hummocks of Ice', 1828, shows how difficult it could be when Parry's crew had to drag their open wooden boats around icebergs and over pressure ridges.

American continent was almost completely unpeopled; only Greenland's Inughuit lived at such high latitudes.

Autonomy is a deceptive concept. To reach the North Pole, Parry's party needed to carry their own food, clothing and fuel. Although he wasn't accompanied by Inuit, he had sought to avail himself of some of the traditional Inuit knowledge he had learned in Igloolik in 1821–2. Parry had taken especially seriously the materials, designs and skills of building and travelling indigenous to the Inuit Arctic. Scoresby too recognized that an amphibious sledge could be designed by putting runners on an Inuit *umiak*, a family or 'women's boat'. An umiak can be carried by half a dozen strong men over a short distance. Parry had evidently made time to study different kinds of sledge design from the many different indigenous cultures around the Arctic. Explaining his plan to readers of *The Times,* Parry extolled for instance the virtues of

George Nares's 1875–6 British Arctic expedition on HMS *Alert* and HMS *Discovery* formalized the dragging of heavily laden sledges as a masculine and heroic test of physical endurance, a technique that became known as 'manhauling'.

the Chukchi *baidar*, a light and shallow, skin-covered coastal craft. Equipping this with a set of metal runners, the craft could be lashed together loosely in the way that Inuit sledges or *kamotiks* were held together, giving them maximum flexibility over uneven terrain. In this way, Parry not only brought indigenous knowledge with him; he and Scoresby had sought out the indigenous technologies necessary for his expedition's mobility in the Arctic.[21]

Parry reckoned that the success of his expedition would boil down to logistics: if he could calculate how many days were required to reach the pole and how much food was necessary per day, correct victualling would more or less ensure success. By the end of the century, Peary would scale up this idea in designing his 'Peary system' of carefully arranged staging posts, each supplied with food rations. Sailing up Smith Sound aboard the *Roosevelt* on his eighth and final expedition, Peary was accompanied by no fewer than 49 Greenlanders and 246 dogs.[22] With this workforce, he was able to set up a forward base at Cape Columbia; from there 'twenty-four men, nineteen sledges, and 133 dogs began the march over the polar ice'.[23] Here then was the logic of industrial organization being applied to polar logistics; the putative autonomy of the polar party was in fact reliant on what was, by polar standards, was a large-scale transportation system of distributing goods.

'The Boats off Walden Island': this sublime scene displays Parry's crews sailing past Walden Island, near Little Table Island. After 48 straight hours in the boats without rest and enduring a gale, the boats took shelter on the shore of what Parry took to be the 'northernmost known land upon the globe' (William Parry, *Narrative of an Attempt to Reach the North Pole* (1828), p. 121).

Scoresby had thought that reindeer would provide excellent pulling power for the sledges. He guessed that on the north coast of Spitsbergen the transplanted reindeer would be able to graze before setting out for the pole. Parry, on the other hand, preferred Inuit huskies to Sami reindeer. While overwintering in Igloolik, he had made a study of the Inuit working their huskies: the number of dogs, the role of the lead dog, the construction of the harnesses, the method of commanding them and so on.[24] He also reasoned that dog teams could be procured more easily from Greenland, where the infrastructure of the Danish colonies was more established. Dogs would add speed to the expedition, and when food was running critically low, they would supply a nutritious source of fresh meat. Here, then, was the template for future pole-bound expeditions, north and south, presaging Peary's use of dogs, as well as the subsequent debates around the relative merits of dogs and ponies in the race to the South Pole.[25]

Worse still for Parry's hopes of attaining the pole was the speed of the ice drifting southwards, sometimes faster than the crew were advancing northwards. So worried was Parry about the impact of the south-drifting ice on the crew's morale that he withheld this information from them, fearing the possible consequences were they to realize that the fight against the ice was being lost. The 24 hours of light played tricks on some crew members, who confessed to having lost track of time. Disoriented, they felt discombobulated, unsure when to sleep.[26] The heightened glare reflecting off the ice during the day brought on snow blindness. To avoid this, they broke camp and travelled by night, even if the difference from day was not immediately discernible to the travellers. It is interesting to see the care with which Parry had prepared for this form of disorientation. He had equipped the expedition with a set of 24-hour chronometers that clearly demarcated the hours of day from night. The risk to the expedition exceeded the physical saturation of the senses. Parry was guarding against the danger that the entire crew, himself included, might mistake night for day and become twelve hours out of sync. Were they to have done so, their astronomical calculations at the pole would very likely have led them in the diametrically opposite direction to the one required, and they would eventually have discovered that they were sailing down the opposite meridian towards Bering Strait and the Pacific Ocean instead of retracing their steps towards home![27]

## Writing the North Pole

Ramping up public expectations before a voyage was easy; making good on those expectations and sustaining them after a failed voyage was much harder. For publishers and planners, failed voyages posed a double problem. Was the voyage in question sufficiently entertaining and instructive to interest a typical enthusiastic reader of travel literature? And could a voyage be phoenix-like, with a new set of goals emerging from the ashes of the one dead in the water? The magnitude of the challenge in the face of repeated failure lent polar conquest an air of impossibility. Very few North Pole narratives were able to recoup their costs in sales. In the second half of the nineteenth century, explorers recognized the need to write popular, and if possible best-selling, narratives.[28] Yarns to entertain readers and raise funds for expeditions became part of the planning of privately funded expeditions. It made sense for them to distance themselves from the dry or technical writing associated with scientific collaborations. Peary's *The North Pole* (1909) was a case in point, in which his expedition to the geographical North Pole was arguably organized around producing a highly profitable display of an American triumph.[29]

The element of human encounter was an essential ingredient in successful travel writing. In the Americas, the territory of northern peoples like the Inuit or the Gwich'in of the Mackenzie Delta reached as far as about 70°N, but not much further; in Greenland, the Inughuit or 'Polar Eskimos' lived as far north as 78°N, offering explorers the opportunity of a uniquely northern inhabited base where a polar party could amass its resources and set out for the pole at a time of its choosing. For other explorers like Parry, setting out from Spitsbergen, indigenous presence in the form of sledge design and travelling techniques was extremely important, but the human drama of indigenous encounters was largely absent.

All this changed when two very different kinds of travellers – Robert Peary, the polar explorer, and Knud Rasmussen (1879–1933), the ethnographer of the Inuit world – set up their exploration bases in the very north of Greenland. Peary brought the Inuit into his expeditionary team; Rasmussen claimed Inuit descent through his maternal grandmother and went one step further, travelling the Inuit world by dog team, in effect indigenizing polar exploration.[30]

North Pole expeditions before Peary failed to make a lasting impact; they struggled to set new 'farthest north' records and were soon forgotten after the event. Readers remember the names of Northwest Passage worthies like John Franklin (1786–1847),

| Year. | POLE N, Strongest Pole in North Hemisphere. | | POLE S, Strongest Pole in South Hemisphere. | | POLE n, Weakest Pole in North Hemisphere. | | POLE s, Weakest Pole in South Hemisphere. | |
|---|---|---|---|---|---|---|---|---|
| | Distance from the North Pole. | Long. WEST from Greenwich. | Distance from the South Pole. | Long. EAST from Greenwich. | Distance from the North Pole. | Long. EAST from Greenwich. | Distance from the South Pole. | Long. WEST from Greenwich. |
| 1800, | 20° 7' | 93° 33' | 20° 53' | 134° 8' | 4° 35' | 131° 43' | 12° 10' | 130° 28' |
| 1810, | 20 15 | 91 28 | 21 1 | 133 21 | 4 42 | 135 54 | 11 57 | 133 14 |
| 1820, | 20 22 | 89 24 | 21 8 | 132 35 | 4 48 | 140 6 | 11 44 | 135 59 |
| 1830, | 20 30 | 87 19 | 21 16 | 131 47 | 4 54 | 144 17 | 11 31 | 137 45 |
| 1840, | 20 38 | 85 15 | 21 23 | 131 1 | 5 0 | 148 28 | 11 19 | 140 31 |
| 1850, | 20 46 | 83 10 | 21 31 | 130 14 | 5 0 | 152 40 | 11 6 | 143 16 |

William Edward Parry (1790–1855), James Clark Ross (1800–1862), Charles Francis Hall (1821–1871), Adolphus Greely (1844–1935), Constantine Phipps (1744–1792), David Buchan (1780–1838) and Frederick William Beechey (1796–1856), but who can remember much about Phipps's or David Buchan's failed North Pole voyages? The icy ocean, repeatedly experienced as drab, monotonous and occasionally very dangerous, made for dreary reading. The hardness of ice ridges was compared with geological formations. The lack of picturesque elements challenged the best writers to produce a stirring composition. When William Beechey eventually published a narrative of his 1818 voyage, he portrayed the ships in a moving sea of giant granite-like ice blocks.[31]

Attempts on the North Pole were therefore sporadic, one-off events until the late nineteenth century. Scoresby talked about polar approach expeditions as 'experiments' because they were in essence speculative trials in which very large machinery and scientific equipment were pitted against a poorly understood, harsh and moving environment. During the months that Parry's sledge parties were battling the ice floes, scientific experiments were performed at the base on Spitsbergen, helping to justify the cost of the whole venture. Even John Barrow was forced to admit that sceptics not invested in the pursuit of pure science or

Christopher Hansteen, more than a century after Halley, claimed to have worked out the trajectories of the Earth's four magnetic poles and, based on past magnetic observations, predicted the poles' future locations (*Brewster's Journal of Science*, 1826).

James Henry Coffin produced a synoptic chart of wind observations in the northern hemisphere (1852), which allowed him to hypothesize the existence of a meterological pole to the north of Greenland.

Plate 1.

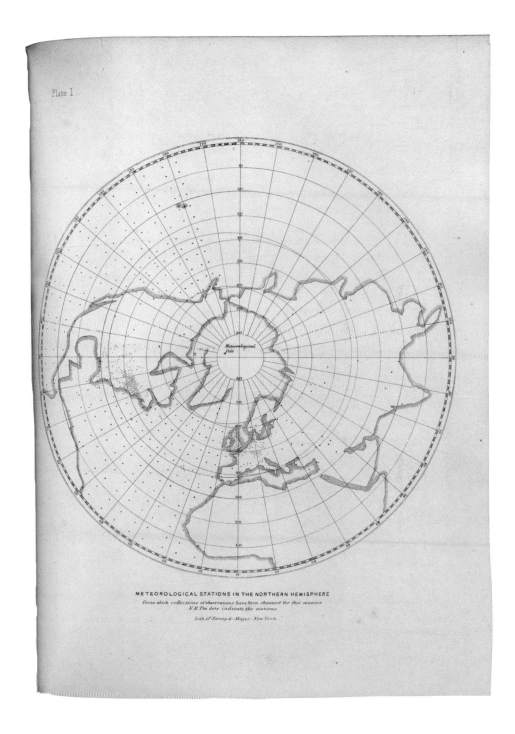

**METEOROLOGICAL STATIONS IN THE NORTHERN HEMISPHERE**

*from which collections of observations have been obtained for this memoir*

*N.B. The dots indicate the stations*

*Lith of Sarony & Major New York.*

geographical knowledge might condemn North Pole approaches as 'absolutely useless' or as 'hopeless and absurd an experiment' as had ever been tried.[32] To justify the expedition to his readers, he returned to the scientific benefits of being in the latitude of Spitsbergen 'nearly equidistant from the two magnetic poles, and from the two cold meridians of the globe'.[33]

A successful narrative depended on officers being able to keep journals and to write on a regular basis. Novel, remarkable, beautiful, dangerous or sublime observations punctuated the mundane accounts of three steps forwards, two steps backwards. But in truth, writing was very difficult in times of prolonged danger when shifting sea and ice threatened expeditions with imminent destruction. If Peary undoubtedly remains the best remembered of North Pole voyagers, David Buchan, commander of the 1818 North Pole expedition, is surely the least remembered. His cardinal mistake was that, while being pummelled by the brutal and crushing force of the polar ice pack, and his ships only just making it back to Spitsbergen for repair, he lost sight of the absolute necessity of keeping a journal. One might think this understandable in the circumstances, but it lost him the trust of John Barrow, the architect of the expeditions, who had established a very close working relationship with the publisher John Murray. This tight-knit alliance possessed the monopoly rights to give them considerable control over the way the voyages would be received and remembered by the public in the voyage narratives and periodical press. When Buchan returned with the commander's written account, there could be no authoritative account of the voyage. It was almost tantamount to the voyage not actually having happened.[34]

Buchan's voyage was not, however, a disaster for all involved. Lieutenant Frederick Beechey, a skilled surveyor and artist, kept a firm grip on his responsibilities and sketched the ships' coastal profiles, the harbours, the scenes of animal life on the shores of Spitsbergen, as well as drawing the ships' charts. He also kept some written notes, if not a full-blown journal. Being such an assiduous recorder served him well. Beechey the artist was soon promoted and remembered, Buchan the commander and failed diarist was shelved. Back in London, Beechey entered into a deal with the proprietor-painter of the Leicester Square Panorama, Henry Aston Barker. He used Beechey's drawings to present to the paying public a truly vast and beautiful Arctic canvas of polar Spitsbergen, 10,000 square feet, larger than anything about the Arctic ever displayed. The scene depicted a scene of providential escape from the ice and deliverance to a safe harbour on the north coast of Spitsbergen. Walruses basked, seals played, polar bears surfaced and birds circled overhead. Word of the dazzling painting, rich in Arctic ecology, quickly spread and the paying public flocked to the panorama to be dazzled by the sensation.[35] Twenty-five years on, when John Barrow wanted all of the voyages he had organized to have proper published narratives, he invited Beechey to pull together all the materials he could muster from his 1818 voyage. In the end it was Beechey's name that became linked to the official narrative, not Buchan's.[36]

Fact and fiction intermingled freely in nineteenth-century popular spectacles. If geographical terminology like 'polar' or 'North Pole' was easily blurred in a respectable newspaper such as *The Times*, promoters designing posters could take liberties in describing their exhibitions as coming from *near* the North Pole, being in its *vicinity*, or simply being *polar*. 'North Pole' or 'polar' began to be used rather loosely to suggest Arctic exoticism. This also marked an emerging trend in which the North Pole began to be used carelessly, perhaps even interchangeably, as shorthand for the Arctic as a whole. With only the writers of Boys' Own adventures and political satirists venturing to describe the North Pole's topography and peoples, who but a few Arctic cognoscenti could easily say in what measure newly collected animal hides, indigenous sledges and other curiosities weren't exactly what one would find in the vicinity of the North Pole? Cheaper printing technologies and popular shows were a licence for proprietors to exploit the geography of the North Pole in creative and lucrative ways that could present the public with remarkable artefacts of indigenous Arctic origin; equally, the easy misuse of labels and provenance was dubious, if not fraudulent, but of little concern to all but the most discerning.[37]

Unable to say what the North Pole was actually like, polar voyages could serve as a bookend to other voyages. Very few North Pole narratives were able to recoup their costs in sales as standalone volumes. However, when presented as a fitting companion to other exploration 'greats', the pole fared much better. Phipps's journal, for example, though published as a standalone volume, was also appended to an abridgement of Cook's voyages and, in another edition, packaged alongside several Northeast Passage expeditions. Parry's 1827 polar narrative became better known as the concluding voyage in the 'connected narrative' that reinforced the image among the reading public that the North Pole was the next chapter following smoothly on from the successive Northwest Passage expeditions (1818–25), serialized in periodicals, moving progressively from one expedition to the next. With publishing costs falling and the market for cheaper abridged books growing, Parry's North Pole voyage became more widely available as the final chapter offered to the public in

The careful attention given to the design of sledges, clothing, and snow and ice tools, from the British Arctic Expedition of 1875–6, epitomized the emergence of polar logistics as a hallmark of polar exploration, a blend of art and science.

smaller and more affordable abridged editions alongside other recent Admiralty Northwest Passage expeditions.[38]

## After the pole

Adolf Nordenskiöld, having traversed the Northeast Passage in the *Vega* (1878), became an assiduous historian of polar cosmography, navigation and cartography, presenting the world views of the 16th-century cosmographers to a new generation of polar audiences.

Retirement after polar voyages troubled many explorers. What work or pastime could be suitable for men who had spent thousands of hours navigating the Arctic ice floes, and who were viewed as a strange species even by other naval surveyors? The quest for Arctic discovery, whether the longed-for Northwest Passage or the experiment of the polar approach, left an indelible mark on the tired spirits and worn bodies of these lionized heroes. Parry and Franklin became colonial administrators in Australia and Tasmania respectively; tiring of the challenges of the then frontier cultures, each returned to London. Franklin was eventually drawn back into Arctic service to renew the quest of discovery; still others begrudgingly returned to the Arctic to

search for lost or stranded expeditions. In one way or another they were enlisted as a group of elder statesmen of the Arctic, men who had shared the unspoken knowledge of having lived in a distant world, experienced privations and found themselves changed, irreversibly, unlike their fellow men at home.

Fridtjof Nansen refined the art of polar skiing while drifting across the polar ocean in the *Fram* (1893–6). Nansen,

The greatest late nineteenth-century explorers of the North Pole – Robert Peary, Adolf Nordenskiöld and Fridtjof Nansen – were intensely fascinated by the mythography that surrounded their quest and their own place in history. Their contemporaries were thinking more deeply about the Earth's long geological past and the origins of their own people in the deep mists of time: Pre-Raphaelites looking back to an age of medieval virtues; archaeologists discovering hunting peoples of the Iron and Bronze Ages; and mythographers celebrating bygone golden ages of lost innocence. The late nineteenth century, however, was also a period of rising imperial competition and tension, and the project of locating the place of human beings in those lost histories took on more importance than ever. Understanding what kind of a past the North Pole might have – geological, human, mythical – became a key factor in placing the polar regions in the grand narrative of history. This new fascination with the past inaugurated the field of 'polar history' as a legacy: an archival exploration into a perceived lineage of navigators, surveyors, cartographers, merchants and monarchs stretching back to ancient times. This field of enquiry may today seem quirky or esoteric, but for many followers it resonates with a certain understanding of the history of the globe itself: a 'polar consciousness'. In this living archive, successive generations of scientific travellers and their chroniclers have been able to look back in time and see themselves as part of a tradition of single-minded and enterprising individuals travelling outside the conventional northern limits of the *oikoumene*.

Thus explorers, accustomed to the requirements of writing their diaries, journals and logbooks, in their retirement became the self-appointed historians of the region whose geography they had mapped. Cartography was the substrate of these polar histories. In his *Facsimile-Atlas to the Early History of Cartography* (1889), Adolf Nordenskiöld compiled and published a remarkable collection of early modern polar cosmographical drawings and charts (discussed in Chapter Three), correctly recognizing the sixteenth-century cosmographers as Europe's most important polar heritage. Nansen trawled through archives and libraries for his *In Northern Mists* (1911), sifting fact from fiction in a study of

like Nordenskiöld, became a dedicated historian of polar navigation in later life.

northern voyages that on closer inspection seemed inexhaustible: the ancients, the Arab scientific travellers, the Norse sagas, right up to his contemporaries. These works travelled across national boundaries with geographical societies supporting translations.

Even at the height of the Heroic Age of exploration, this activity was wedded to the story of cosmography that runs through this book, with all its twists and turns. Polar history was not always the reactionary tradition one might guess would follow from a field steeped in nationalism and mythography. Polar historians, like explorers, could find themselves at odds with the polar images constructed for them by their own country. Nansen found himself embroiled in serious controversy with Norwegian antiquarians when he challenged the literal veracity of the Norse sagas. Capable of distancing himself from his nation's geopolitics when necessary, Nansen went on become a distinguished humanitarian and diplomat, serving as the League of Nations' first High Commissioner for Refugees. The maps of Nordenskjöld's atlas placed the North Pole at the centre, an elevated focal point of a world before nation states, holding in tension the lines of power and navigation of rival empires. The many polar maps in his atlas were a testimony to the skill and ingenuity of artisan craftsmen across Europe in placing cosmography at the service of powerful patrons. The most encyclopaedic and prolific internationalist on polar matters of the twentieth century, Vilhjalmur Stefansson, was called before the U.S. Senate Subcommittee on Internal Security in 1951 on the grounds that his interest in the Russian Arctic belied communist political beliefs.

The North Pole could, of course, serve reactionary political ideologies. As one might expect, polar historians of high empire could be highly nationalistic and self-serving. Clements Markham, like John Barrow before him, showed only begrudging international respect and goodwill towards colleagues from other countries. He was very clearly aware that, as President of the Royal Geographical Society, he was at the helm of a tradition of science and exploration in the service of the British Empire, and Markham's many works in historical geography made heroic the construction of Britain's imperial archive. In the next chapter, we take a closer look at polar utopias and satires, where we find some of the most radical and sinister uses of the North Pole to give credence to utopian ideologies, and importantly those people who challenged their dogmas: satirists and critics.

# 5 Polar Edens

Voyages to the poles were of course made by practical and intrepid seafarers, but they were also custom-made for utopians, mystics and fantasists. Not only was it no disadvantage for mystics not to have been to a pole, at least not physically transported there; it positively helped that the North Pole, the actual place, was unknown to human eyes. If no one with authority could reliably describe the poles, who could definitively say that the wildest and most extravagant claims were untrue? Safe from prying eyes, the Earth's polar axis and poles possessed a strong appeal as places for locating narratives and symbols of absolute sacredness and purity. Evidence of these in sacred texts lured very different kinds of philosophers of metaphysics, including those we might term as esoteric or belonging to the occult. These philosophers found in these texts endlessly fascinating opportunities to interpret mysterious clues and patterns underlying what they discerned to be deep hidden truths about the origins of the globe and its earliest peoples.

Helena Petrovna Blavatsky (1831–1891), the founder of the Theosophical Society, and René Guénon (1886–1951), a student of Hinduism and Islamic esoteric philosophy, were two of the most influential esotericists to make forceful arguments about the mystical importance of the geographical North Pole.[1] Blavatsky, a Russian-German aristocrat, emigrated to the United States in 1873, before moving on to India in 1880. Theosophy promised to reveal an ancient wisdom on which rested all of the world's religions. Blavatsky was frequently dismissed as a fraudulent

Helena Blavatsky (1831–1891), aristocrat, mystic and co-founder of the Theosophical Society in 1875, was both revered by followers and reviled by many as a fraud. She looked to esoteric ideas in Neoplatonism and Hermeticism for inspiration for a unified vision of science, religion and philosophy.

spirit medium, but theosophy spread widely throughout India. Guénon, who took on the style of an intellectual, wrote against theosophy, criticizing it as superficial and tainted by British imperialism. By contrast, he claimed to be an active member or participant in an 'initiatic science' of esoteric knowledge, unlike those who merely scratched the surface. Guénon insisted that Hyperborea had been an ancient continent situated at the North Pole. Sacred Mount Meru symbolized the polar axis, aligning the poles with the *Arktoi* constellations of Ursa Major and Ursa Minor, echoing aspects of ancient cosmography.[2] For metaphysical philosophers like Guénon and Blavatsky, chasing after the place to study its material existence was of secondary importance. For example, Blavatsky's 'First Race' at the North Pole was 'purely ethereal', existing without material form.[3] In spite of the distinctly different schools and approaches, the many strands of mystic and occult philosophies when braided together gave the North Pole a powerful symbolic significance.

In earlier chapters the idea of the polar axis was closely linked to ideas of an Eden, a terrestrial paradise, a place of perfection or divine virtue. For some early modern cosmographers, such as Oronce Finé, the North Pole was thought to be the seat of paradise. For others the pole was a mere vestige, a pale shadow of the divinity preserved in the celestial heavens. This polar Eden, once a garden paradise, had been forfeited in Christianity by original sin as told in the book of Genesis. The possibility that the North Pole had once possessed a warm and bountiful climate persisted in many forms in the centuries that followed. Geographical utopias of an open polar sea, which persisted into the late nineteenth century, were consistent with the idea of an Edenic pole. Even the belief that the North Pole lay at the heart of a cold, icy region could be reconciled with the state of Eden after the human fall from grace. Explanations of climatic change from a much warmer period could be made to support these views rather than dispel them. A shift of the polar axis, for example, could explain the presence of fossils in high latitudes. As is often the case, evidence to dispel myths can be reworked to lend support to them. The idea that providence placed an obligation on humanity to make the world agriculturally productive and to restore the lost state of Eden was a powerful narrative that continued to shape European thinking about climate and landscape throughout modern times.[4]

Religious speculators debated where the Hyperborean descendants of those forced out of the Polar Eden might have migrated to and rebuilt their world. Theories suggesting sites across Greenland, northern Asia or island archipelagos like Atlantis submerged in the North Atlantic Ocean all attracted followers.[5] Occult traditions held out many such geographical

Oronce Finé's cordiform (heart-shaped) map of the world, 1536, watercolour engraving.

origins as possibilities, but what they shared in common was the supposition that they could lift the veil on this polar Eden, and unlock the secrets of its past shrouded in the hidden mysteries of Hindu, Persian, biblical and classical texts.

The poles were openings to rich sources of inspiration and possibility beyond the surfaces of the inhabited world. Mount Meru, described in the sacred Hindu texts, the Puranas, is the centre of Hyperborea resting on four supports made of gold, iron, silver and brass, each pointing to one of the cardinal points of the compass. The river Ganges is of celestial origin, issuing from 'the feet of Vishnu near the Pole Star', some 1,080,000 kilometres (672,000 mi.) above the Earth. Thus the North Pole, at some considerable altitude, is the physical location of the source of life.

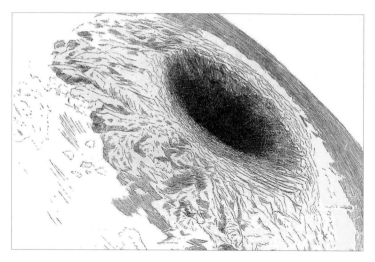

John Cleves Symmes's image of a hole in the North Pole as seen by a Lunarian through a telescope, from *Harper's New Monthly Magazine* (1882).

For Madame Blavatsky, the continent Hyberborea occupied a different immaterial and less literal metaphysical space and therefore couldn't be grasped or claimed by imperial conquest. Even if present at the North Pole in physical space, one could witness nothing of Hyperborea because it existed in another parallel space. This attested to its power being a source of truth of a kind at once strange and beyond normal reach. Thus nineteenth-century popular writers like Jules Verne or Edgar Allan Poe could draw their readers into stories of the dark side of the North Pole and its weird otherworldliness, ascending upwards or descending into the Earth's interior space, variously described as a passage boiling, demonic or hollow.[6]

For many readers, whether early modern or late twentieth century, the boundary between works of fact and fiction was often blurred. What lay above or beneath the pole was more tantalizing than the surface impressions observed, measured and tabulated as useful information by writers of scientific travel and exploration narratives. Human beings have for most of their history been anatomically and technologically confined to occupying a thin band of space along the surface of the Earth (before the invention of submarines and aircraft). The power of human imagination and desire to explore the far reaches, nooks and crannies of the Earth through myths, rituals and storytelling has always run ahead of what scientific travellers witnessed. In the ancient world, wax-winged Icarus took to flight to gain his freedom, flying higher and higher, ever closer to the Sun, until he discovered his mortality, falling to his death.[7] Aeneas, fleeing the ruins of Troy, journeyed to the shores of the underworld, passing through its shadows en route to founding Rome.[8] Each of these stories contained lessons about hubris and empire. Icarus and Aeneas both made moral passages crossing thresholds and entering into other worlds above and below the Earth: the upper limit of the atmosphere for Icarus, the river Styx as the gateway to the underworld for Aeneas. Each journey involved a transformation that made the possibility of return fraught with danger.

So, too, the North Pole as a literary or narrative device was a gate or threshold that warned off those who approached, a signpost that travelling beyond the pole was irrevocable, an undertaking with no prospect of returning unchanged. This was a condition of passing into a world of a different kind, the one separate, incompatible and irreducible to the other. The Jesuit philosopher Athanasius Kircher (1602–1680), citing medieval sources in his *Mundus subterraneus* (1665), described the North Pole as being covered by a very large black rock, 33 leagues in circumference, beneath which four sea channels drained into the polar sea. Directly beneath the pole itself, a whirlpool was said to drain downwards through the Earth's interior, its currents emerging at the South Pole. Kircher likened these polar vortices to the circulation of blood in animals or humans, giving life to the body and preventing the poles from freezing. The animus of these powerful currents could also explain why the North Pole was so difficult for navigators to attain. The Arctic navigator Henry Hudson (1565–1611) met his end at the hands of a mutinous crew, one could argue, precisely because of the obstacle of flowing ice and fast-moving ocean currents that resisted his efforts to sail northwards.[9] Navigating a safe course through these fast-moving currents required understanding of the behaviour and spirits moving them, something only acquired with constant practice and long experience – of which only the Inuit could claim to have a deep knowledge.

# Le Petit Parisien

## SUPPLÉMENT LITTÉRAIRE ILLUSTRÉ

TOUS LES JOURS
Le Petit Parisien
(six pages)
5 centimes

CHAQUE SEMAINE
LE SUPPLÉMENT LITTÉRAIRE
5 centimes

**DIRECTION: 18, rue d'Enghien (10ᵉ), PARIS**

ABONNEMENTS

PARIS ET DÉPARTEMENTS :
12 mois, 4 fr. 50. 6 mois, 2 fr. 25
UNION POSTALE :
12 mois, 5 fr. 50. 6 mois, 3 fr.

## LES HOMMES VOLANTS
### Les Ailes d'Icare et l'Aéroplane d'Henri Farman

Athanasius Kircher's image of the Earth's interior was inspired by a study of Mount Vesuvius, which, after climbing, he lowered himself into to observe the extraordinary interaction of air, fire and water in its crater (1678).

Icarus used as an allegory for the beginnings of aviation, on the cover of the December 1907 issue of *Le Petit Parisien*.

The North Pole was far too important to be the sole preserve of male chroniclers, seafarers and mystics. Margaret Cavendish (1623–1673), Duchess of Newcastle, published what is possibly the most remarkable early modern narrative of polar exploration. *The Description of a New World, called the Blazing-world* (1666) was published the year after Kircher's treatise.[10] Her husband, William Cavendish, had been a commander of the Royalist army at Marston Moor in 1644 and then went into exile, marrying Margaret in Paris in 1645. They were both able to return to England in 1660. Margaret was an erudite noblewoman whom some feminists have claimed as an important proto-feminist. Breaking with convention, Cavendish had deeply immersed herself in debates on natural philosophy and the nature of matter. By writing in this genre of fiction and making difficult theory entertaining, Cavendish was also making a statement about her

literary identity, that she, as a woman, could become an author in a public way of complex and erudite philosophical ideas.[11]

    *The Blazing World* is set in a very strange polar utopia in which Cavendish herself seems to possess a dual character. It opens with her narrating the story of her crossing the polar threshold. Her journey to the North Pole was not of her own making. Rather,

Margaret Cavendish (1623–1673) depicted beside her desk: engraving after Abraham van Diepenbeeck, 1800.

she was kidnapped by a suitor and taken prisoner. Their ship is blown off course northwards to the polar regions, where the crew encounter cold so intense that it derives not merely from one pole, but two, 'so that the cold having a double strength at the conjunction of those two Poles, was insupportable'.[12] The feeble crew perish, but the lady does not on account of her ardent spirit! The duchess, being a vitalist about the North Pole, makes it a place in which spirit exerts a powerful force on matter, and her boat is propelled into an alternative world joined to ours at the poles. Cavendish explains how the worlds are connected thus:

> For it is impossible to round this Worlds [*sic*] Globe
> from Pole to Pole, so as we do from East to West;
> because the Poles of the other World, joining to the
> Poles of this, do not allow any further passage to
> surround the World that way; but if any one arrives
> to either of these Poles, he is either forced to return,
> or to enter into another World.[13]

Entry into other worlds takes place via the North Pole, a gateway through which those who pass discover transformed worlds. For Cavendish these worlds were places of encounter with other species, each with its own political culture and hierarchies. Her protagonist travels through a succession of other worlds inhabited by strange races of animals with their own morals and customs. Eventually she arrives at the last of these worlds, where she meets the Empress who becomes her ruler; she in turn, the author of this strange tale, becomes her Empress's scribe. In its own way, *The Blazing World* is a story of an inverted world, radically reoriented by the author's own transformation beyond the pole. Is the Empress in fact Cavendish's other self, as some have argued? Power in this world is gendered female, not male. Liberty, sexuality and identity burst the social constraints of Restoration Britain and worlds are revealed in which Cavendish pays homage to greater empires of other genders and species.[14]

Scholars have looked back on Cavendish with admiration for her spirit: independent, learned and libertine. One hundred and fifty years later, in 1816, the young Mary Shelley (daughter of the most celebrated British advocate of women's rights, Mary Wollstonecraft), wrote an equally self-conscious work of fiction, her first draft of the Romantic novel *Frankenstein*, in which the eponymous doctor builds a human being by applying an electrical spark in a life-giving experiment in galvanic chemistry. Frankenstein becomes tormented however, realizing that he is incapable of reciprocating the love his new creation needs. Like those in *The Blazing World*, Shelley's creation is a powerful being, neither wholly human nor beast, inspiring fear and perceived as monstrous by his parent and creator, Dr Frankenstein, who turns his back on his creation. The opening and climax of the book take the form of a great chase on the sea ice of the frozen polar sea. The encounter between Dr Frankenstein and Captain Walton takes on a homoerotic tone when Captain Walton sees a growing fraternal love in Frankenstein's eyes as he lays in his cabin recuperating. Walton's intention to discover the 'secrets of the magnet' parallels Frankenstein's search for the spark of life in chemistry.[15]

Beneath the surface of this Gothic romance, Mary Shelley and her husband, the poet Percy Bysshe Shelley (1792–1822), were much taken with radical scientific materialism, the idea that matter itself might contain life force (rather than a spirit). By the religious standards of this day, this looked rather too close to atheism. Ice was a source of particular fascination for Romantic poets and artists, associated with the stillness of death and therefore a purity of spirit that might in some way be generative of life. The poets Robert Southey (1774–1843) and James Montgomery (1771–1854) and the painter Caspar David Friedrich (1774–1840) were each in very different ways smitten by the associations of ice with death and rebirth. Percy Shelley, too, had been much taken by the romantic chemistry of Humphry Davy (1778–1829), as well as the physiology espoused by his personal physician, William Lawrence (1783–1867), who argued that matter itself could have agency and be self-organizing. Frankenstein's declaration that the source of desire to possess the North Pole is no different than that of burning to possess the secret key to the origin of life is revealing. The quest for absolute knowledge, biological or geographical, is a perilous one, tragic for both creator and creation.

Crucially for Cavendish and Shelley, polarity is a central principle in their respective narratives. Earthly poles are the surface points where powerful interior forces of magnetism burst through the surface, and galvanic poles represent the surface points where inner seas of oppositely charged chemical particles have affinities of either attraction or repulsion. These life-giving poles, whether magnetic, chemical or cosmographical, leave no room for compromise and brook no dissent. The characters and the social orders, even when inverted as in *The Blazing World*, gravitate towards extreme hierarchies in pursuit of self-understanding. This seems to be the dominant, even defining, feature of polar worlds throughout the passage of time.

In Samuel Taylor Coleridge's famous poem 'The Rime of the Ancient Mariner' (1798), the narrator approaching the southern Antarctic continent encounters the 'Spirit of the Pole', akin to a classical goddess. This almost certainly played some role in inspiring later polar spirit narratives, including the satire *Munchausen at the Pole* (1819) and Edgar Allan Poe's *The Wonderful Adventures of Arthur Gordon Pym of Nantucket* (1838).[16] Romanticism thus contributed allegories in polar thinking looking back to Gilbert's experimental studies of ensouled magnetism, and even further to the philosophy of the Neoplatonists. The Romantics' enduring drive to contemplate an inner self in the face of nature, pushed to its extremes, was concerned with the relationship between life and death.

## Polar satire

Uniting satire and Romanticism in their concerns about polar exploration was the shared danger of hubris. While the poles could serve to inspire, they were also ripe for satire and political dissent. The British Admiralty invested so much state authority and prestige in its early nineteenth-century mapping of the globe's polar regions that poking fun at the state and puncturing its pretensions became too tempting a target to resist.[17] In the hands of journalists and satirists, the poles and the reputations of those seeking them could be made comical or farcical.

The heroic polar expeditions of Ross, Parry and Franklin were certainly ripe for the picking of satirists. The reactionary government of Lord Liverpool (1812–27) was especially anti-democratic, stemming from the fear of a French invasion during the Napoleonic Wars. The British state clamped down hard on dissent. Church and state fuelled a burning hatred of the threat from France as ungodly, revolutionary and expansionist. In the years following Napoleon's capitulation at Waterloo, British radicals and democrats like the journalists William Cobbett and William Hazlitt lived dangerous lives and faced a constant threat of persecution. Speaking truth to power could take many forms, but to do so risked the wrath of the Tory administration and the long reach of its network of reactionary officials and spies.[18] The Admiralty's architect of polar exploration, John Barrow, was always minded to track down critics and even to cut down to size perfectly innocent authors whose published views just happened to oppose his.[19]

Opponents of this anti-Jacobin government could see the Royal Navy's polar voyages as a form of nationalist propaganda. Arguing against polar exploration was difficult when at moments it was so sensationalist. Experienced mariners could and did argue that government exploration strategies were flawed and offered better plans in their place, but it was left to satirists to suggest that voyages were largely pointless and badly planned. With a good satirical cartoon, the public might at one moment enjoy a good joke at the government's expense, just as at another moment it might cheer on the occasional discovery or sighting of a 'furthest north'. Behind the satire grew a realization that the struggle to sail, saw and pull large ships across the surface of large fields of ice revealed the absurdity of the human condition being unable to escape the limitations of inhabiting the surface of the Earth.[20]

Simply announcing the intention to make a North Pole voyage could liven up the House of Commons no end. This is the sort of theatre that Captain Thomas Cochrane MP (1775–1860) seemed to have in mind in 1818, when he rose to address the House.

A PATRIOT LUMINARY EXTINGUISHING NOXIOUS GAS !!!

Satirical etching by George Cruikshank of the politician Henry Brougham dousing Thomas Cochrane aboard a vessel with the French insignia, 1817.

Today he is remembered through the fictional characters Horatio Hornblower and Captain Jack Aubrey, the latter brought to life as the protagonist of the film *Master and Commander* (2003) and played by Russell Crowe. Cochrane was simultaneously a would-be polar conqueror and a larger-than-life showman and satirist. This naval hero was a veteran of the Napoleonic Wars, a radical democrat, an aristocrat, a Member of Parliament and bête noire of the conservative establishment. When Cochrane announced to the House of Commons that he would personally liberate Chile from the Spanish by sailing a ship to Chile, he proposed – almost as an afterthought – sailing via the North Pole. By taking this very British route instead of rounding the Spanish-dominated Cape Horn, Cochrane could claim the Longitude Prize along the way! Anyone doubting that this maverick hero wouldn't do just what he said could visit Rotherhithe where the ship *North Pole* was being converted to a steam-powered warship rechristened the *Rising Star*![21]

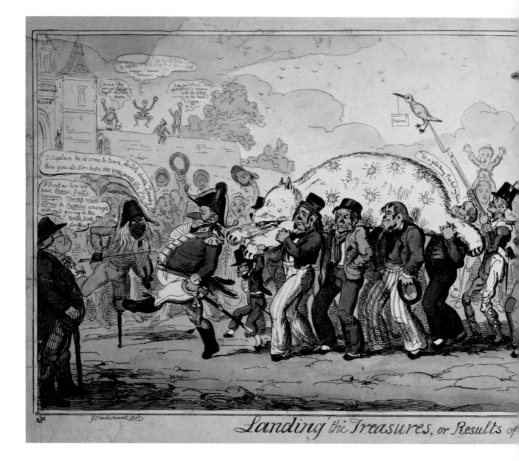

Landing the Treasures, or Results of

Had Cochrane seriously entertained bagging the North Pole on his way to Chile? No one in the House of Commons that day could have mistaken the anti-state, point-scoring in Cochrane's oratory. The difficulty in distinguishing fact from fiction was in no small measure what made Cochrane so appealing. His polar exploit never materialized, but the voyage to liberate Chile certainly did, and the polar speech became a matter of parliamentary record.[22]

The North Pole itself was satirized for the reason that the quest to discover it encapsulated the values of state hauteur and vanity. After the Ross and Buchan northern expeditions of 1818 had spectacularly failed to deliver on the Navy's overblown

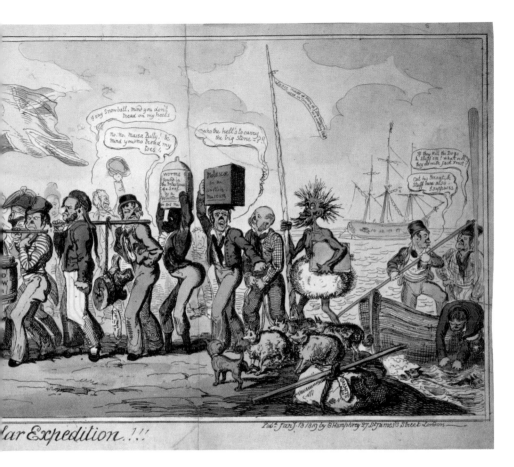

George Cruikshank, 'Landing the Treasures, or Results of the Polar Expedition!!!', 1819, etching.

promise to find either a polar or northwest passage through the Arctic, it was open season for satirists. A George Cruikshank print of 1819, 'Landing the Treasures, or Results of the Polar Expedition!!!', showed the Ross expedition marching home with a ragbag retinue of marvels including Jack Frost, a polar bear shot through with bullet holes in the shape of Ursa Major, and other ridiculous curiosities. When Captain John Ross and his nephew, Lieutenant James Clark Ross, returned to London in 1835 after locating the North Magnetic Pole, they became the object of Cruikshank's wit in his engraving of Midshipman Blockhead's 'Arrival at the North [Magnetic] Pole'. More mirthful than political, Blockhead was shown climbing an extremely tall pole, by then a familiar way of punning the geographical and magnetic poles.[23]

*Munchausen at the Pole* (1819) was likely the most relentless and revealing political satire in the history of polar exploration, mocking the credentials and ambitions of the Ross and Buchan expeditions. The story contains a stream of unmistakable cultural references and asides, and is shot through with polar jokes as well as broadsides at Regency contemporaries. Although various Munchausen stories had been published for a couple of decades, the character of this Munchausen bears more than a passing resemblance in style to Thomas Cochrane.

The story opens with the 'spark of ambition' of the bombastic and illiberal Baron Munchausen, being 'revived afresh at the mention of [the Longitude Act's] £20,000 reward'.[24] The path to the pole through Baffin Bay is blocked by vast icebergs beneath which are the 'living tomb of the Titans', who defend the sacred territory from the incursion of outsiders. The Titans struggle to free themselves to 'assist the spirit of the Pole, who presides over and protects the sacred magnet, in expelling me by terror from her dominions, which tradition had truly told her would one day acknowledge a mortal's power'.[25] Munchausen navigates these dangers thanks to being guided first by a magical serpent that

George Cruikshank's 'Arrival at the North Magnetic Pole in 1831', from his series of satirical prints of the progress of Midshipman Blockhead.

A scene showing expedition members celebrating as they watch James Clark Ross raising the British flag on the North Magnetic Pole (one of two purported), 1 June 1831. In his diary, Ross describes the pole as unremarkable, producing a lingering feeling of melancholy.

swims the Northwest Passage through to the Bering Strait, and then by the goddess Chance northwards to the pole. There the Spirit of the pole defends a place that resembles Hell, and threatens Munchausen with vengeance if he climbs the giant Pole or 'axletree' of the universe – which of course he does.

When Munchausen climbs the North Pole, he reaches a platform where he finds an old jewel-shaped shack with an inscription on the door in an unearthly language that says 'seek wisdom and she will be found of thee'. True to form, Munchausen shatters the door with a blow from his sabre. On the table inside, he finds a golden book called *The History of Science*. Seizing it, he opens the sacred text and discovers that every page is blank![26] He then resumes his ascent and climbs over a 'magnetic cross which had no effect on me', which leads him to arrive at the top some thousand miles above the Earth. He gazes down on the Sun, Moon, stars and kingdoms of the world and registers his triumph by raising a Union Jack and proclaiming George III to be 'Monarch of the Polar Regions'.[27] Greek Titans, spirits, magnets, the celestial axis and flags – all the key ingredients in the formulation of polar imperial power are present in this account. From a thousand miles above the Earth,

'Munchausen Entering the Capital of East Greenland', from *Munchausen at the Pole* (1819).

'Munchausen
Shaved and Lathered
with Red Snow'.
A volcano is seen
in the background.

Munchausen observes with self-satisfaction that 'the prospect below me was inexpressibly grand. The sun, moon, and stars, revolved beneath, or remained stationary, and . . . before me lay all the kingdoms of the world.'[28]

Is the sacred book at the North Pole, the *History of Science*, a symbol of polar exploration's false promises, the emptiness of the claim to reveal any true science, or perhaps a history of science yet to be written? That science holds the keys to the secrets of the pole is revealed to be a shallow conceit, and the supposedly scientific arguments are either hidden or vacuous. This accusation chimed with other critics sceptical of Barrow's claim that providential intervention was causing the Arctic climate to warm suddenly and for the thick ice barrier surrounding the open polar sea to melt, break up into large icebergs and to float south with the sea currents. In the eyes of many politically as well as scientifically informed critics, satire was an alternative to reasoned argument, and in some respects more appealing because it offered the most direct way of challenging the state's pretensions to cloak exploration with the reputation of respectable science, and used the language of nationalist propaganda to show it for what it really was.[29]

The joke about the North Pole being an axle pole with platforms for viewing the world reminds us that the quest for the geographical pole was never simply about a very remote place on the Earth's surface. Since early modern times, the pole had always been about hidden forces, celestial harmonies, viewing the Earth from the heavens or transformations deep inside the Earth's interior. Well into the nineteenth century, writing about the North Pole remained steeped in the past ideas of the Earth's place in the universe: the early modern cosmography of the Ptolemaic universe, the magnetical philosophy of Gilbert, the mechanical philosophy of Halley. These historical layers remained important because they helped the public make sense of a proliferation of poles that the Enlightenment could neither master through voyaging nor reasonably explain with science.

Munchausen surrounded by Africa, America (background) and Europe (foreground) proclaims possession for George III of 'all Countries upon and beyond the Pole'.

The British state insisted that polar exploration should fly the flag of nationalism and take itself seriously, by being empirical, disciplined and serious in its pursuit of science, but readers were at least as entertained by mystery, encounter and the strange stories of Inuit shamanism.[30] Serious or not, the Arctic mattered to Britons because cosmography and natural philosophy had instilled in ordinary people a belief that the Arctic made the globe whole in a very British way.

Visual culture played a critical role in shaping how the Victorian public imagined the Arctic. Museum exhibitions, pantomimes, panoramas, magazines, lantern slides and other forms of popular entertainment suffused a vivid visual vocabulary of the Arctic as never before. The rise of mass media brought about new technologies that made Arctic imagery feel ubiquitous. This was also a period in which increasingly rigid distinctions about gender circulated and polar landscapes became seen as feminized. When the Nares expedition to the North Pole (1875–6) was turned back with several men having died of scurvy, the North Pole was personified in *Punch* as a frigid and inhospitable 'Ice Queen' whom the crews had failed to make submissive. This reflected a growing eroticism in exploration and use of female imagery to represent nature that emerged with the doomed Franklin expedition (1845–8) and became more pronounced as the suffering of the rebuffed expeditions became more known.[31] The use sometimes of the masculine 'old King, Jack Frost' for the North Pole portrayed the resistance of a despot, in which the expedition's failure was framed more as a military setback after fighting against stiff odds.[32] The very significant role played by Greenlandic families, men and women, often absent or misrepresented in satire, were actually recorded on camera during the Nares expedition, one of the earliest to use photography during a polar approach.[33]

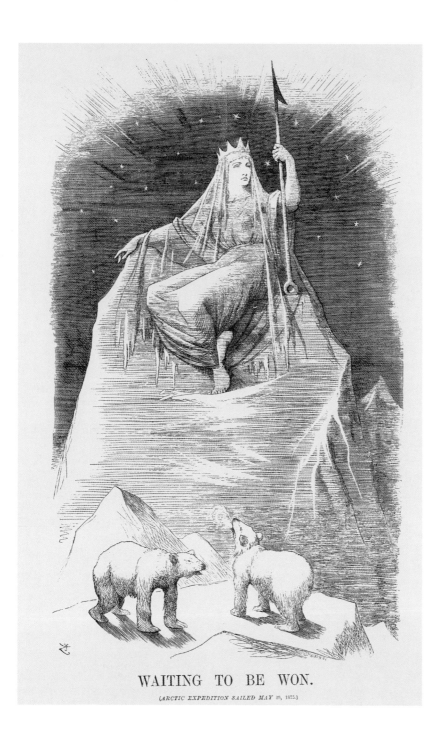

## WAITING TO BE WON.

(*ARCTIC EXPEDITION SAILED MAY 29, 1875.*)

## Victorian polar historians

Had science and exploration abolished polar mythology in the wake of enlightened exploration, one would not be surprised. But the reverse happened; polar archetypes of paradise, race and origins came back with a vengeance in the second half of the nineteenth century. New historical disciplines in the human sciences like archaeology and anthropology, together with geology, were oriented to shed light on humanity's origins and deep past. Polar explorers, historians and antiquarians began to lift the lid on the early voyages of intrepid Arctic navigators. As these scholars pored over evidence from manuscripts, fossils and archaeological artefacts, they began to piece together fragments of the migrations to show how the North Pole played a central role in the early story of humanity. Not always distilling fact from myth, it also became the case that facts were used to sustain myths to support grand theories. As we shall see, this had great consequence for millions of innocent people who had no desire whatsoever to lay claim to the poles.

The great Victorian polar explorers felt a deep-seated need in their later years to place their polar journeys and their contributions in the larger story of humanity discovering the poles through the march of time. Polar heroes of Norway, Sweden and the United States – Fridtjof Nansen, Adolf Erik Nordenskiöld and Robert Peary – each in distinctive ways became historians of the polar quest. Nansen threw himself into the task of lifting the veil on the Arctic Eden to connect himself with the long lineage of polar navigators, particularly Pytheas, the fourth-century BCE navigator from Marseilles, an independent Greek city in a trading alliance with Rome. Pytheas was an important political figure in the late nineteenth century. Lionized by Napoleon III, his statue adorned the new stock exchange in his home city.[34] Historians of northern countries combed the fragments of his periplus for evidence to argue that Pytheas' Thule corresponded to a location in modern-day Norway, Iceland or Shetland.[35]

Nordenskiöld contributed to polar history by devoting himself to the history of early modern cartography. He discovered

'Waiting to be Won: The Queen of the Arctic is Perched', *Punch* (5 June 1875).

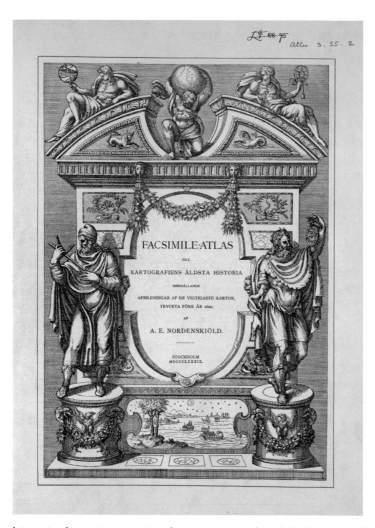

The astrolabe, globe, compass and blackstaff adorning the frontispiece to Nordenskiöld's atlas reveal the importance of navigation and cosmography to his history of cartography.

historical inspiration in the cosmographical brilliance of cartographers like Werner, Finé and Mercator. Maps on a polar projection formed a very substantial part of his *Facsimile Atlas* (1889), in a period when the history of cartography was itself a new subject of study.[36] The Royal Geographical Society began to act as a clearing house in this period, facilitating the translation of historical geography from other languages into English. To that end Nordenskiöld teamed up with Clements Markham, the English doyen of imperial and classical geography, a veteran of George Nares's pole-bound British Arctic Expedition (1875–6) and Secretary of the Royal Geographical Society (1863–88).

Robert Peary, too, ardently embraced an imperial vision of historical geography, at once classical and nationalist in its foundations. He likened himself to a latter-day Herakles, destined to vanquish the ancient Titan god of the North Pole.[37] To dismiss the polar histories of any of these explorers as amateurish retirement hobbies would underestimate them and miss the

A marble statue of Pytheas adorns the Marseille Bourse, a proud emblem of the city's civic and commercial values.

point. These distinguished men of science were fully intent on telling the story of the true nature of the polar Eden, building on the knowledge of the voyages of early modern navigators and their predecessors from the ancient world. This tradition of historical scholarship was reshaping the subject of geography for an imperial age in which European nationalism placed the northern hemisphere front and centre in world history, educating its young men with a mythography and curriculum of national progress and civic virtue.[38] The North Pole was therefore a central part of the story of imperial aspiration for circumpolar nations, but that was not all.

The deep mythology or belief in the sanctity of a polar Eden had travelled far beyond the borders of the circumpolar world. This history had, it seemed, travelled out of Eden with the people who had themselves been forced out by cataclysmic events. Piecing this story together was no easy task. William Warren (1833–1929), Boston University's first president and a professor of comparative religion, argued, against the tide of contemporary evolutionary theory, that at the North Pole there

Mercator placed his 'Arcticus Polus' at the centre of an 'Ulterior incognita' in the cordiform projection of the two hemispheres (1538, repro. 1889).

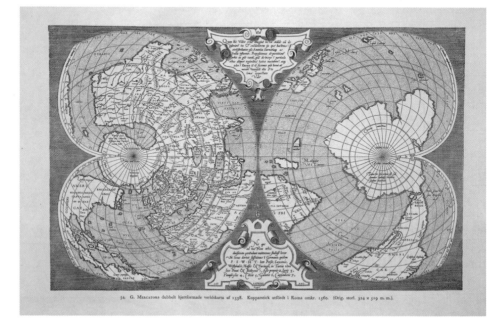

54. G. Mercators dubbelt hjertformade verldskarta af 1538. Kopparstick utfördt i Roma omkr. 1560. (Orig. storl. 324 x 519 m. m.).

Georg von Rosen,
*Adolf Nordenskiold*,
1886, oil on canvas.
With the *Vega* in
the background,
Nordenskiöld's
striking pose on the
sea ice illustrates the
virtues of courage,
determination and
curiosity.

was an 'antediluvian continent', the original site of paradise and
nothing less than the 'cradle of the human race'.[39] This great
Edenic crucible of civilization had been left submerged first by
the biblical deluge and then beneath a great advancing ice sheet
brought about by a shift in the Earth's polar axis and a sudden
cooling of the circumpolar climate. The fleet-footed survivors of
this calamity moved quickly south from the ruins of Eden to
establish the earliest Aryan settlements in northern Central Asia.

Others believed they had settled in Greenland, or northern Scandinavia. Though this polar theory will seem bizarre to many readers today, Warren was no crackpot working in isolation. Variations of his ideas about Aryanism were far from uncommon, and his own influence on others was considerable. Indeed Josceyln Godwin, a historian of the polar archetype, offers a sober warning that these occult readings have a very substantial following today among an educated, literate audience who should know better.[40]

Warren claimed that the sacred paradise with its centre at the North Pole was nothing less than a 'primeval circumpolar mother region'. This circumpolar land was 'divided into four nearly equal quarters'[41] joined to the celestial circumpolar sky by an extremely tall mountain from whose rivers the Earth receives its waters. In one sense, Warren's polar history was a return to

This portrait of Peary, aboard the *Roosevelt*, shows the close relationship he had with the dogs he relied upon for his travel over the polar sea ice.

early modern cosmographical utopias of the sixteenth century, such as those of Oronce Finé and Gerard Mercator (see Chapter Two). In another sense his circumpolar history was distinctly modern, presented as a compendium of evidence sifted by the new scientific disciplines of archaeology, anthropology, geology and linguistics, using texts from the world's ancient religions, selectively, to muster evidence to validate this sacred cosmography. What made his work so distinctive was the prominence of a comparative method, gathering in evidence beyond a single nation or religion. Thus his evidence situating polar paradise at

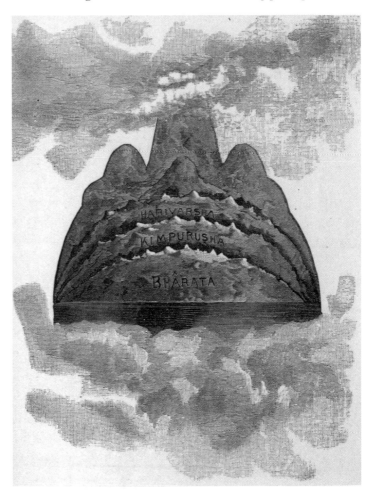

'The Earth of the Hindus', from William Warren, *Paradise Found* (1885).

the 'navel of the earth' was derived from Iranian, Chinese, Japanese, Buddhist, Hindu and Greek oral or sacred texts. The evidence in these texts – at least Warren's use of them – converged on his trans-religious hypothesis.

The political significance of polar Aryanism lacked a unified voice, but it appealed to different kinds of radical nationalists. Taking inspiration from Warren's *Paradise Found* (1885), one of India's most prominent nationalist and pro-independence leaders, Bal Gangadhar Tilak (1856–1920), published his own sacred history, *The Arctic Home in the Vedas* (1903). Tilak, an astronomer by training, had become a political journalist and leader, preaching radical resistance against British colonial occupation.[42] Hindu Vedic sources were valuable to Hindu nationalists because of their religious importance, and their links to ancient Hindu sciences gave them considerable prestige. Tilak used his astronomical training to trace the origins of Hindu and Aryan peoples back to an interglacial period (*c.* 8000 BCE) when the climate of the Arctic had been much more temperate, enabling him to construct a history where the epochs of Aryan migrations culminated in the logical need for a political expression of Hindu nationalism.

The term 'circumpolar' remains poignant today if only because it has important political resonances of cooperation and internationalism. Warren was identifying something else, a region with a polar universalist cosmology, a common racial and geographical heritage – but whose heritage was this? Mythical theories of Aryan origins located in a great sweep from Arctic Scandinavia or Asia southwards into Central Asia and Europe, and between India over to Persia and Europe, held a potent appeal to ethnonationalists in many societies. The use of astronomy to lend scientific credence to such ideas had been growing since the eighteenth century in the work of important astronomers like the French revolutionary Jean Sylvain Bailly (1736–1793) and the German geographer Johann Gottfried Herder (1744–1803). Famously a small group of German National Socialists around Rudolf Hess, for example, formed the Thule Gesellschaft to explore the occult associations of Aryan lineages. For National Socialism, the polar origins served as a repudiation

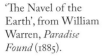

'The Navel of the Earth', from William Warren, *Paradise Found* (1885).

of the traditional orientation of geography towards the sacred sites of the Judaic Mediterranean.

Polar ethnonationalism and imperialism were difficult to resist, but satire and mockery continued to play the lead role, as Cruikshank and other political cartoonists had done a century earlier in an age of reactionary politics. The pages of *Punch* and the stages of London's East End music halls were quick to satirize Peary's purported polar conquest. In *Punch*, for instance, the then assistant editor and later dramatist and children's author A. A. Milne (1882–1956) published a poem, 'An Unconvincing Narrative' (1909), in which an expedition led by a warm-hearted cockney had reached the 'Great Big Nail' of a North Pole and removed it before Peary and Cook arrived.[43] The verse, besides mocking their temporal priority to discovery, also made fun of their rigid racial and social hierarchies. Milne's expedition in verse comprised a crew of several well-known sportsmen of the day, as well as an Inuk (Etukishook, one of Frederick Cook's two Inuit co-travellers), led by a working-class man. How far Milne's audience bought into the joke is impossible to gauge, but the poem belongs to a tradition of satirical polar counter-narrative.[44]

In *Winnie-the-Pooh* (1926), written in the wake of the First World War, Christopher Robin leads an 'expotition [*sic*] to the

North Pole'. Milne had acquired a philosophical understanding of logic while reading mathematics at Cambridge. When Piglet asks Pooh what they are going to discover, he replies, 'Oh! Just something', and when asked whether it was fierce, Pooh answers, 'Christopher Robin didn't say anything about fierce. He just said it had an "x".'[45]

When Kanga's baby Roo falls in the water, he is rescued by Pooh with the help of a long Pole he's picked up. When asked where he found the pole, Pooh declares that he 'just found it' as he 'thought it ought to be useful'. Christopher Robin then solemnly declares that the pole is the North Pole, and that 'the Expedition is over'.[46] The manner of Pooh's discovery can be read as a nostalgic yearning for a lost innocence, and the game is ended when the North Pole is revealed to be something at hand and mundane. The mixture of anxiety and bemusement of the expedition party registers the ambivalence of the author towards the Boys' Own stories that glorify imperial adventure, and a degree of scepticism about the North Pole, identifying it as a variable 'x'.

Milne, diverging from ethnonationalists who elevated the status of the North Pole to that of an *ur*-site of Aryan origins, recognized it for what it was, the essential point of origin in a mathematical projection, but philosophically no more special than anywhere else. Hence Milne's philosophy of mathematics served to buttress his anti-imperialism, views that only strengthened in the years to follow, such that in the 1930s Milne took up the cause of pacifism.[47]

# 6 Sovereigns of the Pole

Historians have for at least a generation cast doubt on the notion that the triumph of science and rationality in Enlightenment Europe really caused a decline in the belief in magic.[1] As states, aided by science and technology, have adopted new strategies for rational models of governing their resources, nature has resisted attempts to take possession of it. The North Pole certainly resisted its would-be conquerors by virtue of the difficulty of getting there; its inaccessibility being the product of a fierce desert-like climate and a scarcity of food that demanded expeditions carry all they need.

In the face of such a ferocious climate, how did myths of Edenic nature around the North Pole so powerfully persist into the twentieth century? The idea of an open polar sea, championed by at least two of the most persuasive geographers in Europe, John Barrow and August Petermann, had a history reaching back to theories about the formation of ice in the early seventeenth century, if not earlier.[2] Curiously, as successive explorers or approachers established new 'furthest north' records and put paid to the open polar sea theory, the North Pole retained its ethereal aura of enchantment. One reason is that being a polar explorer was still genuinely hazardous. Expeditions like Vilhjalmur Stefansson's Canadian Arctic Expedition (1913–16) could end in shipwreck and tragedy even after a century of exploration of Canada's archipelago, never mind seeking the North Pole. The technological revolution in exploration brought new ways of circumventing the ice – aerial and underwater – but they also brought new dangers, and North Pole approaches still ruthlessly claimed lives. For example, five members of Umberto Nobile's 1928 expedition perished when they were 'carried off to their deaths by the wrecked but still airborne *Italia* airship's envelope and keel'.[3] Search expeditions were prone to being equally obsessive and dangerous.[4] In the ensuing international search effort to rescue Nobile's shipwrecked party, the aircraft carrying Roald Amundsen, his pilot René Guilbaud and four others went down over the sea between Tromsø and Svalbard, killing all on board.[5] The Arctic Ocean was more navigable in new ways, but it remained sublimely untamed without a reliable source of food on the drifting sea ice.

The desire to possess the North Pole, though driven by a kind of masculinity found in an era of extraordinary nationalism, cannot be reduced to blind ambition or folly. As we saw in Chapter Five, explorers thirsted for a self-understanding of their place in a sacred history. One dimension of this was a story of enchantment, the lifting of the veil on the former Eden in the guise of an Atlantis or paradise that had been scattered and lost by a great flood or the onset of glaciation. The other dimension was a story of knowledge and progress, seeing themselves as belonging to a lineage of explorers who successively and cumulatively were advancing geographical knowledge towards its full maturity. While it is possible to see these two historically minded dimensions as being in tension or contradictory, they were both present in the polar visions embraced by explorers of the age.

North Pole visionaries at the turn of the twentieth century were prophets of their day seeking an ultimate truth. Some were would-be emperors and courtiers to world leaders wielding substantial political power, while others were solitary figures speaking from the shadowy margins of society. Like the figure of the court fool in the eighteenth century, a fine line divided the reputation of visionaries and jesters.[6] So, too, it was with the North Pole, this most special of geographical places: defined in its purest form as simply a location in relation to celestial measurements, and yet hardly a place at all, even a non-place. With this ambiguity, it is easy to see how this timeless point on the polar axis attracted utopians, Machiavellian schemers and satirists in equal measure.

The ambiguity of the visionary and the mad fool was beautifully portrayed in a story of polar political intrigue by Jules Verne. *The Purchase of the North Pole* (1891) resonates with the story of nation states today vying to claim the right to own the resources at the North Pole.[7] Dystopian machinations around greed, profit and resources abounded, then as now, in the name of responsible frontier capitalism. The story opens with an announcement that the North Pole and all the territory extending down to the 84th parallel is being put on the auction block in Baltimore. A consortium of anonymous private investors called the North Polar Practical Association (NPPA) conspires to purchase 'an indefeasible title to all the continents, islands, inlets, rocks, seas, lakes, rivers, and watercourses whatsoever of which this Arctic territory is composed, although these may be now covered with ice, which ice may in summertime disappear'.[8] Their aim is to have title to this very desirable slice of land and sea, equal in area to about 'nearly a tenth of the whole of Europe – a good-sized estate!'[9] They seek the legal rights to develop the polar territory, allegedly to pre-empt the possibility that the United States might give the unclaimed territory to someone else. The premise is reminiscent of the Russian sale of Alaska to the United States in 1867 for 1 cent per acre. The Russians were motivated to pre-empt the possibility of Britain seizing Alaska in a future war. In the United States the purchase, though popular, was dubbed by sceptics 'Seward's folly' after Secretary of State William H. Seward.[10]

In Verne's story, the NPPA scheme is initially viewed by the public as having originated 'in the brain of a fool'. Bidding opens at 10 cents per square mile among the representatives of the key polar states: Russia, Britain, Denmark and Sweden – the latter is represented by Jan Harald, Professor of Cosmography in Christiania (now Oslo – Norway was then under Swedish rule). The price rises quickly until Major Donellan, the British representative, bids 100 cents per square mile and the other parties back down. Only then does Mr Forster, the NPPA's agent,

'Any advance on forty cents?' George Roux's illustration of the North Pole for sale on the auction block, from Jules Verne, *The Purchase of the North Pole* (1891).

and who has until then been silent, suddenly join the fray and enter into a bidding war. Forster pushes Donellan to the limit of his government's purse, and the hammer goes down after he bids an unbelievable 200 cents per square mile, silencing the room and leaving Donellan overwhelmed.[11] In due course, however, it is revealed that the funder behind the NPPA is none other than a gun club, Barbicane & Co., and its mastermind, the redoubtable astronomer J. T. Maston. His reputation as a visionary genius and a 'brilliant calculator [of] all mathematical formulae' precedes him, figuring in Verne's earlier story about a voyage to the Moon. Maston is a Renaissance man and Victorian cosmographer, and so skilful were his beautifully drawn Greek letters that 'an Archimedes might have been well proud of them'.[12]

By no coincidence, an astronomer and a firearms magnate are Verne's polar villains. Pulling all the strings, J. T. Maston devotes himself to the most difficult calculations in mechanics to produce a master plan of geoengineering. His secret purpose is to access the hidden coalfields of the Arctic. The public dismisses this possibility on account of the high cost of accessing coal deep beneath the ice sheet, requiring technology that doesn't yet exist. Maston, however, calculates that an explosive blast from a cannon perfectly calibrated using his complex equations will shift the Earth's axis precisely the right amount, thereby changing the climate at the North Pole. As the tilted pole moves southwards, the warmer temperatures will melt the polar ice, exposing the rich veins of coal. The resulting fall in the cost of extraction will make the evil Maston and his shady investors untold wealth!

For Victorian readers, the themes of reclaiming a lost Eden, a rapid climate change caused by a tilting polar axis, and private capital funds working in the background were all familiar. Verne's protagonist understood that a warming climate could suit the short-term interests of speculators. The nefarious hijacking of a new Eden or Atlantis in the guise of coalfields spoke to readers aware of rampant financial speculation in commodities. To this Verne added an important dimension, a scientific rationale for the technological control over Arctic nature on the grounds of restoring the region to the fecund fauna of a former geological epoch. The large coalfields found at the North Pole, Barbicane explains, originated 'in the carboniferous period ... [when] immense forests covered the northern regions long before the appearance of man'. This theory 'the journals, reviews, and magazines that supported the North Polar Practical Association insisted on in a thousand articles, popular and scientific'. Major Donellan, a diplomat wise to resource geopolitics, had to concede that 'there will

'The Making of a New Axis was Possible', from Jules Verne's *The Purchase of the North Pole* (1891).

be fortunes made in exploring this region' because 'North America has immense deposits of coal . . . [and] the Arctic regions seem to be a part of the American continent geologically, similar in formation and physiography . . . and certainly Greenland belongs to America'.[13] Donellan observes that Adolf Nordenskiöld,

> when he explored Greenland, found among the sandstones and schists intercalations of lignite with many forest plants.

Even in the Disko district, Steenstrup [a Danish authority
on Greenlandic geology] discovered eleven localities with
abundant vestiges of the luxuriant vegetation which formerly
encircled the Pole.[14]

Thus the NPPA's cunning plan turned on a vision of the Arctic as
formerly having a warmer and fecund climate that could be cyn-
ically re-engineered by a small elite of self-interested industrialists
and investors.

In fact, securing leases on polar minerals and hydrocarbons
at a deflated price with the intention of marshalling technology
to make extraction more accessible and profitable came to pass
not long after Verne published *The Purchase of the North Pole*.
After the First World War polar veterans of Antarctic exploration
became involved in supporting speculative schemes for mineral
prospecting in the Svalbard archipelago. A proposal to annex
Svalbard (then Spitsbergen) to assert sovereign rights over its
resources actually circulated in Britain and won the backing of
a significant group of Fellows at the Royal Geographical Society
but was ultimately blocked by the Foreign Office.[15]

Verne answers the question of whether the protagonist
Maston is more genius or fool by incorporating satire into the
story:

Here was a chance for the caricaturists! In the windows
of the shops and kiosks of the great cities of Europe and
America there appeared thousands of sketches and prints
displaying Impey Barbicane . . . driving a sub-marine tunnel
through masses of ice, which was to emerge at the very
point of the axis . . . Here J. T. Maston, who was as popular
as Barbicane with the caricaturists, had been seized by
the magnetic attraction of the Pole, and was fast held
to the ground by his metal hook.[16]

Verne thus settles on a precarious balance between the plans of
powerful speculators and an unbelieving public that imagines it
knows better.

## Peary and Herakles

Verne's polar contemporaries, the hardened explorers of the Golden Age of polar exploration – Peary, Nansen, Amundsen, Nordenskiöld, Scott, Shackleton – were all, in their own ways, visionaries. While it is true that each individual desperately wished to claim the honour for the 'conquest of the pole' before the others, theirs was a deeply personal quest to endure and bear witness to something metaphysical, an inner calling. As with most visionaries, the purity of their ambitions could be called into question: were they gifted with a capacity to see beyond where ordinary mortals can, or were they self-deluded in hawking hollow promises to gullible punters? By extension, was laying claim to the North Pole an extraordinary feat of bearing witness to the most sacred of earthly places on the globe, or was it more akin to the *History of Science*'s blank pages discovered by Munchausen at the North Pole (see Chapter Five)? There is no simple answer to this: neither possibility can be entirely dismissed.

For these *fin de siècle* explorers, the lure of spiritual self-knowledge pointed to some notion of polar interiority. Although the idioms and styles of this inward journey differed among individual explorers, a common thread ran throughout them. The fundamental condition for inner self-discovery they shared was to make a journey to a gateway through which they might pass to transcend earthly space and be spiritually transformed. The North Pole was never just about arriving. It entailed crossing a boundary to enter an unmapped world of a different kind. From the time of Halley, philosophers and mariners had mused about sailing 'beyond the pole' or 'into the pole'.[17] The hardship and moral hazards pointed to an uncertain space: who could say with certainty whether the conquest of the pole was metaphysically a real place or a door into other possible worlds?

Peary, Nordenskiöld and Nansen all viewed the epic voyages of classical heroes as their ancestry and inheritance. Like the eighteenth-century noblemen who designed their princely gardens around very carefully chosen busts of their classical heroes, so too polar explorers wrote themselves into a lineage of navigators

'Hatteras Climbs the Volcano to Plant an English Flag at the North Pole but Falls into an Abyss', from Jules Verne's *The Adventures of Captain Hatteras* (1874).

whose fortunes were made and lost by the quarrels of jealous gods.[18] Mythical histories were also very practical when it came to fundraising to finance expeditions. Selling outlandish hopes, building reputations, raising subscriptions, filling lecture halls, designing broadsheets and feeding newspaper editors all required a dab hand and feel for myth-making. In the early twentieth century newspapers and journalists in Britain and the United States began to promote and popularize polar exploration as never before.[19]

Robert Peary, credited by many as being the first person to set foot near the North Pole (1909), likened himself to a modern-day

Herakles. Putting to one side his long and bitter controversy with Dr Frederick Cook over priority of discovery – the reliability and veracity of their respective tracks, survey instruments and astronomical records were thrown into doubt – Peary's own portrayal of his imperious self was striking. While fundraising in 1906 for his final bid to reach the pole, he wrote an article for the *Pall Mall Gazette* likening himself to Herakles, writing himself into a lineage of heroic imperial predecessors.

How Peary portrayed the North Pole as his adversary was still more significant: as Antaeus, a Titan god, giant, imperious and unforgiving, a vast bearded body rising up out of the polar ice towards the celestial realm, ominous and threatening to

Cook's carefully constructed portrait at the North Pole shows him as a man of science, harnessing animal and nature, with a rather unrealistic portrayal of Inuit companions in the background.

trespassers. With the foreground littered with the corpses of failed expeditions, the stage was set to depict Peary's polar trek as an epic battle of mythical proportions.[20]

Gaia, living Mother Earth, is also revealed here in what is arguably one of the most evocative and surprising pieces of Peary's polar writing: 'Though her ribs are gaunt and protruding with the cold and starvation of centuries, nowhere else does one get so close to the great heart of Mother Earth as up there in that borderland between this world and interstellar space which we call the Arctic Regions.'[21] This borderland between Mother Earth and interstellar space makes sense of the physical geography of the journey by recourse to classical myth. Peary replies, 'there is to be found the realisation of the fable of Antaeus', the son of Gaia and Poseidon, the gods of the earth and the oceans. Herakles encounters Antaeus while seeking the golden apples of the Hesperides, which he has to bring back. After defeating Antaeus, Herakles meets Atlas, who is holding the celestial heavens on his shoulders, and offers to relieve him of his burden temporarily if Atlas will get the apples for him. Atlas does this,

This postcard of 1909 celebrating Cook and Peary placing the American flag on the polar axis subsumes their mutual animosity beneath a banner of nationalism and empire.

TWO DAUNTLESS AMERICANS WHO REACHED THE GOAL OF A THOUSAND YEARS AND PLANTED THE STARS AND STRIPES UPON THE AXIS OF THE WORLD.

but doesn't want to take back the burden of holding the heavens, requiring Herakles to trick him into doing so.

Antaeus is a formidable and deadly wrestler who takes on all comers. According to Plutarch, after challenging and killing those unfortunate enough to cross his path, Antaeus uses their skulls to build a temple in homage to his father Poseidon. The polar version with the twelve labours (tasks or challenges) of Herakles is clear: those who seek the pole will face a violent end at the hands of merciless Antaeus, as the bodies strewn around the sledges in the foreground of the drawing testify.

Herakles fighting Antaeus depicted on a shallow bowl, from the workshop of Giorgio Andreoli, 1520.

The story of Herakles and Antaeus had been popular among Renaissance artists such as the Florentine brothers Antonio and Piero Pollaiuolo, whose studies of anatomy informed painted and sculptural depictions of this theme commissioned by Lorenzo de' Medici around 1475. This battle to the death attracted artists into the twentieth century and has been retold in popular writing and compilations of classical myths. The Pollaiuolo brothers' fascination with anatomical study offers a subtle clue to the polar

Marcantonio Raimondi after Raphael, 'Hercules and Antaeus', after 1520, engraving.

conquest version of the myth. In Antonio's bronze of the subject, Antaeus is rendered as organic, alive. His torso is one with the ice fields, his oesophagus is the shape of a tall spiralling pole that seems to draw breath into, or perhaps from, Mother Earth. His secret is that the source of his strength and power is his physical contact with his mother, Gaia, such that if he is thrown to the ground or simply falls, he is instantly healed and renewed. After Herakles accepts Antaeus' challenge and they start to wrestle, he senses how Antaeus falls to the ground slightly prematurely, before being actually thrown down. Herakles then grasps Antaeus' secret, and so lifts him off the ground, suspending him in the air and so weakening him. By this way Antaeus is defeated. The idea that the North Pole is organic, that it has a natural history, a spirit not to be trifled with, invites us to think more critically about the pole as a holistic, spiritual and grounded place. It also reveals Peary to be a vitalist, someone who, like the utopians discussed in Chapter Five, should make readers cautious about the dangers of polar romanticism when combined with trenchant ideas and hierarchies of race.

The relationship between Antaeus and Herakles was from the outset a story of empire. Herakles (known to the Romans as Hercules) was famously an Olympian, a man of enormous strength and courage. The Olympians in Greek mythology were a new order who had overthrown the Titans. Antaeus' violent wrestling contests have been interpreted as a symbol of resistance against the new imperial power of the Olympians, and Herakles as the Olympian who quashes this Titan rebellion. The efficacy of the myth is that the story of the combatants can be told from more than one perspective and that its protagonists, individually or jointly, can be read in different ways. For example Theodorus, who debates with Socrates in Plato's *Sophist*, describes Socrates as an Antaeus and emphasizes his insistence on always contesting arguments, grounding them as dialogues, contests in moral philosophy.[22] If Antaeus is Theodorus' Socrates, then Peary could be seen to be Thrasymachus, the strong-headed protagonist in Plato's *Republic*, who argues that surely justice should serve the interests of the strong.

GADITANAS COLVMNAS STATVIT
HERCVLES.

Hans Sebald Beham, 'Hercules and the Columns of Gaza', from *The Labours of Hercules* set of engravings, 1545. Carrying the Columns of Gaza to set up his Pillars was one of the labours of Hercules assigned to him by Hera.

Peary respects Antaeus' polar mystical purity, as this embodies the spiritual alterity of his true adversary who resists his expeditionary advances with Socratic wisdom and a powerful and unrelenting stubbornness. This North Pole is the undiluted realm of nature alive in itself. 'Nowhere else' on Earth, Peary writes, 'is the air so pure . . . the sunlight so brilliant, or the darkness so opaque . . . the storms so furious'.[23] 'The great day' and the 'great night' of the polar year, visions of standing beneath 'Polaris in the very centre overhead', are a testimony to the awe and intimacy of a landscape he knows best from a dozen years inhabiting the Arctic, in fulfilment of an ambition for a military conquest that means he can never be at home there. It is a vision of a particular heroic code of masculinity that seeks purity at the margins of society above all else, where there is a nature rid of all refinement, comfort and excess; and it speaks to the nineteenth-century Romantics' fascination with ice, death and rebirth.[24]

Peary's depiction of Antaeus, albeit a fragment in his writing, fascinates in part because it makes public what the reader takes

a mixture of measurement and guesswork under inauspicious conditions. Where to plant flags was more a matter of leaving a trail and judging local conditions than producing undisputed astronomical calculations, contrary to what J. T. Maston claims in *The Purchase of the North Pole*. What signalled Peary's metaphysical presence at the pole was the sense that he had travelled beyond it, having 'passed over or very near the point where north and south and east and west blend into one'.[27]

Peary's imperial vision looking southwards over the surrounding globe is reminiscent of the dream of Scipio Aemilianus, who had been adopted by the elder son of Scipio Africanus, hero of the Second Punic War (218–201 BCE). Conquerors, past and future, soar together over the Earth and, gazing down over Carthage, presage the younger Scipio's destruction of it. In that moment the elder Scipio sees that the prize of Carthage and in fact the entire Roman Empire is insignificant when compared with the perfection of the celestial heavens. It is a feature of cosmography that the North Pole, once regarded as an insignificant point on the axis of the universe beneath the celestial pole, is elevated to the sublime in Peary's published narrative, in which he describes it as 'the precise centre of the northern hemisphere, the hemisphere of land, of population, of civilization . . . and the last great geographical prize' that the world has 'to offer to adventurous man'.[28] The vision bears the imprint of Peary's nationalism. On the journey to the North Pole, he wraps the Stars and Stripes around his body, cutting away segments of fabric at each camp as a ritual act of territorial possession. In that sense the allegory of Scipio's dream reinforces the vision of a dominant American (Roman) empire, dwarfed only by the divinity of the celestial realm.

In 'The Lure of the North Pole' (1906), while preparing for his expedition, Peary invites his readers to imagine witnessing his 'roll call', as he calls forth his team at the top of the world. Whereas 'for nearly a thousand years, Norseman and Dane, Briton, German and American, have crept painfully northward . . . [here] were gathered the representatives of three great races – myself, the Caucasian, Henson the Ethiopian, Anghmaloktok the [Inuit]

Mongolian' – in that racial order. 'Then there are the dogs, four of them, members of my own team – the "Old Guard" as I called them'.[29] Taking inspiration from the Roman Empire, a federation of peoples with distant polities, united in paying tribute to the emperor, Peary's order announces itself through a construction of race as the hierarchy of the global order.

Three years later, on the final leg of his journey to the pole, Peary's instruments revealed that his party had overshot and gone 'beyond the pole'; this was understandable, as he had no easy way of judging if or when he had attained the desired latitude. On this 'march of only a few hours', he reflected that 'on the first miles ... we had been travelling due north, while, on the last few miles of the same march, we had been travelling south, although we had all the time been travelling precisely in the same direction'.[30] What did this paradox of the reversal of orientation

Peary selected this stellar projection to illustrate his vision of the North Pole as the conjuncture of the continents and peoples of the globe. Taken from Peary, *The North Pole* (1910).

amount to? Peary's response is revealing. Acknowledging that it 'would be difficult to imagine a better illustration of the fact that most things [including time] are relative', he also saw in this timeless place a cosmographical vision reminiscent of Odysseus, Alexander or Aeneas: 'I had passed from the western to the eastern hemisphere and had verified my position at the summit of the world.'[31] Thus the imperial gaze from the North Pole, looking southward, from the zenith of the politically dominant northern hemisphere, is registered as a colonizing passage through which he passes in an instant out of the West and into the Orient.

Antaeus portrayed as the personification of the North Pole, embodied and emerging from the ice, with icy vaulted arches a temple to his father Poseidon, possibly also depicting the northern lights. From the *Pall Mall Gazette*: 'The Lure of the North Pole'.

Matthew Alexander Henson (1866–1955), polar explorer. Henson's remarkable life story and appreciation of Inuit culture unsettle the heroic classical mythology that Peary so strongly embraced.

If Antaeus is the defeated sovereign at the centre of the globe's empires, the Stars and Stripes that Peary drapes himself in and plants on the hastily built summit at the pole are themselves sovereign to no one. The United States becomes the lawgiver that governs the navigation and commerce of the world's civilizations, even if simultaneously Peary's mythology is at odds with the reality that President Taft declined the opportunity to claim sovereignty over the North Pole. It is, however, a general feature

Hand-coloured photograph of Peary's North Pole party showing Matthew Henson with the Stars and Stripes (centre). Flanking him are four Inuit: on his right are Ooqueah and Ootah holding Peary's Navy Leage and Delta Kappa Epsilon fraternity flags. On his left are Egingwah and Seegloo holding Peary's Daughters of the American Revolution Peace flag and Red Cross flag, 1910.

of emperors and monarchs that they seek a universal authority for their legal authority, but are not themselves subject to those laws. That universalism has an equivalent polar cosmography. Peary expounds to his readers a paradox that becomes a defining trope of the pole's mystery for modern audiences: 'There is no time at the North Pole . . . there are no meridians, or rather all the meridians of the globe are gathered in one point, so that there is no starting-point for time as we estimate it here.'[32] The pole is timeless; and yet it is exactly because the Earth rotates around its polar axis that there exist meridians (or lines of longitude) that come to provide a standardized measure of time and space for scientific navigation, maritime trade and international standards.

This brings us to an important crossroads in reflecting on the embodiment of the North Pole as the son of Gaia. If it is a cautionary tale about the need to remain grounded, it is at heart a story that is also firmly rooted in the labours of Herakles, which was always an imperial mythography through and through. Peary was writing at a time of intense international imperial rivalry. Britain, the United States and Germany were all acquiring unprecedented military sea power. Fridtjof Nansen was spearheading Norway's mapping of the polar basin. Salomon Andrée's 1897 expedition to the North Pole by balloon was an experiment in charting an aerial passage that would be followed only thirty years later by trans-polar aviation routes.

Nowhere is the story of early twentieth-century polar cartography better told than in the artwork of MacDonald 'Max' Gill. His two hand-painted domed cupolas (1934) for the museum of the Scott Polar Research Institute in Cambridge, England, represent twin maps of the polar regions, testifying to the lineage of polar worthies who contributed to their exploration from the time of Pytheas. Gill had a profound understanding of both the symbolic and practical value of polar cartography. In those same years, he was commissioned to produce circumpolar drawings to celebrate the British Empire's global-scale post, telegraph and shipping routes. His artworks remind us that the polar sovereignty was less about laying claim to the ownership of the

"THE PROFESSOR HAD CAPTURED THE POLE"

"RAPIDLY MELTING UNDER A HOT SUN"

The icy North Pole is captured, only to melt under the heat of the Sun, in these satirical drawings of the Muggsen Expedition. From *The Ludgate* (August 1896), pp. 379–84.

The artist of the Arctic dome (1934) of the Scott Polar Research Institute, MacDonald Gill, portrayed a historical narrative of polar exploration beginning with early modern ships venturing into the sea ice. Gill frequently used polar projections in this period to portray the enormous global span of the British Empire using themes of telegraphy, shipping and post, as well as exploration.

North Pole, than it was a means of displaying the control of access to imperial networks of science, communications and trade. How the story of Antaeus might be read through the lens of other non-imperial or postcolonial allegories is a task taken up in the final chapter.

# 7  Mourning Antaeus

Over the course of the twentieth century, the colonization of the Arctic Ocean took place on an entirely unprecedented scale. A key question addressed in this chapter is whether the arrival of industrial technology, particularly aviation, would rob the North Pole of its mythical status and the magic surrounding the Arctic. Inasmuch as aviation 'promised omniscience and omnipotence', Marionne Cronin has observed that the 'pilot in his cockpit far above the icy surface seemed to subvert the heroism of polar travel'.[1]

Dethroning the poles had been welcomed by some. Robert Rudmose-Brown (1879–1957), who was Professor of Geography at the University of Sheffield, a renowned botanist and a veteran of Bruce's Scottish National Antarctic Expedition,[2] was pleased that 'the conquest of the poles has allowed exploration to be divested into more useful directions than the mere attainment of a high latitude'. Writing in 1933, Rudmose-Brown had seen how aviation had begun to transform polar science, bringing surveys a previously unimaginable range and access to polar landscapes. At the same time he also regretted the downside, that 'it has robbed polar work of a popular zest, and will not make it easy to raise funds for a large-scale expedition in the future'.[3] His ambivalence about the demise of the poles' mythical status was shared among the increasingly professionalized community of polar scientists, albeit in different ways. Had the North Pole's star set, and had it become just another icy point on the map? Or would it remain an enchanted, distinctive and exceptional place, a touchstone for reflecting on humanity's relationship to the cosmos?

In the stories and perspectives that follow, it will become apparent that a lively geographical imagination of the North Pole persisted against the odds. Arctic work and travel were rooted in 'technological imaginaries' in which flight and aerial instruments spoke to the geographical imagination. For readers of exploration to picture an aeroplane or submarine crossing the Arctic Ocean required giving them a visual context and structure for showing them how to view the polar world from above. On a distant day, horizons might stretch for a hundred miles in each direction. Approaching the North Pole, gazing down over it, and travelling beyond it had been part of a geographical imagination for at least four centuries, but the naturalism that cameras could bring from high above the Arctic Ocean was something new.

The older values of polar travel as recorded by explorers in their journals and books remained very popular. Young polar scientists, notably in the Oxford and Cambridge expeditions of the 1930s, remained interested in seeking out Herculean feats of masculine endurance: for example, the Cambridge University expedition that crossed the Greenland ice cap with a party of only two with little backup.[4] There was also a changing of the guard, something different from their predecessors about the post-1918 generation. Gino Watkins (1907–1932), post-war expedition leader to Greenland and expert kayaker, was forward-looking. He preferred 'men of no polar experience, unless that experience had been with him . . . so as to break with effete traditions . . . and encourage new technique'. His expedition members possessed 'an appetite for risk-taking and rugged, physical endeavour, [that] had not diminished'.[5] For Watkins, ideas about masculinity played a defining role in how polar men conceived of their own identities and their sense of intergenerational change in polar research.

As the North Pole became incorporated into global networks of maritime and air transport, civilian and military, two overarching themes became more pronounced in mid-twentieth-century geography: technology and accessibility. Whether Peary had actually reached the North Pole in 1909 divided public opinion, but it was nevertheless widely accepted as a genuine achievement of some kind by polar contemporaries – including Roald Amundsen, who recognized that his window to be first to the North Pole had closed and turned his sights on the South Pole instead. The 'conquest' of the North Pole ruptured the perception that it was an isolated place, even if Peary's achievement was eventually relegated to the ambiguous status of a 'furthest north'. At stake in the debates over the status of Peary's achievement were successive moments of enchantment and disenchantment in an age when the mechanization of polar transport was acquiring a certain aura or charisma.[6]

How people thought about the potential of mechanical technologies to transform the world became irrevocably altered by the First World War. With the mechanization of warfare and the loss of nearly an entire generation of young men, the Victorian promise of Promethean technologies to bring progress was viewed with far greater ambivalence.

## Stefansson's Pole of Inaccessibility

One person who was particularly concerned by the implications of technology for the future of the Arctic was Vilhjalmur Stefansson (1879–1962), veteran explorer and philosopher of all manner of Arctic subjects. He readily acknowledged that Peary had reached the North Pole, or near enough, in 1909. After the First World War he also readily accepted that it could only be a matter of time before aeroplanes would be flying transpolar routes between major cities in Europe and North America. Stefansson regretted that the conquest of the North Pole had inevitably led to a loss of enchantment, just as the 'Greeks drove their gods off Olympus through the perverse scaling of the mountain to its top'.[7] Stefansson wanted the northern edge of the world to stand for something beyond a mere race or a dash to a nondescript pole. He wanted to restore the tradition that beyond the geographical North Pole was a sacrosanct inner space. How he did so reveals a lot about polar geography in the twentieth century.

In his writing, Stefansson moved beyond the North Pole's surface appearance, literally and metaphorically, by explaining to his readers what was hidden beneath the pole. What if the North Pole were celebrated for the values and way of life found in the Arctic itself, rather than by images of conquering the environment? Thus Stefansson set out on a philosophical journey to reflect on the conventional meaning ascribed to the geographical

A rugged-looking Vilhjalmur Stefansson, foremost polymath, naturalist and popularizer of the Arctic in the United States, is shown here in the Arctic summer during the Canadian Arctic Expedition (1913–16).

North Pole – and to move beyond it. Historians have remarked on the irony and wit that infused Stefansson's writing and the lightness of touch with which he expressed his most serious ideas.[8] True to his ironic style, he theorized into existence a new hidden pole, the so-called 'Pole of Inaccessibility'.[9]

By the early 1920s anthropologists had begun to show Western audiences the world as seen through the eyes of indigenous

Peary, in this frontispiece to *The North Pole* (1910), depicted himself in his 'Actual North Pole Costume' of caribou and polar bear clothing with an Inuit spear and Dene snowshoes, suggesting a close identification with indigenous technologies of travel, rather than the strongly nationalistic vision he otherwise championed.

peoples themselves. In the Pacific, Bronisław Malinowski published his seminal account of four years living among the Trobriand Islanders.[10] In the Arctic Robert Flaherty released his innovative visual documentary portrait of Inuit life at a small camp in Hudson Bay.[11] Only one year earlier Stefansson had published *The Friendly Arctic* (1921), a groundbreaking study of Inuit life and their culture of mutual dependence on animals for food, clothing and spiritual sustenance.[12]

Stefansson's book reflected his belief that the Arctic frontier was accessible to exploration and economic development if one took the trouble to learn how to live in harmony with the Arctic; whereas struggling against the Arctic environment, as though it were unforgiving and harsh, rendered that world inaccessible. The philosophical way forward was to acquire less commonsense knowledge rather than more, and to 'forget what we think we already know', particularly when it came to technology and exploration.[13]

*The Friendly Arctic* was devoted to the time Stefansson spent living with Inuit, in 1913–17, in the delta of the Mackenzie River, where it flows into the Arctic Ocean – nearly 2,500 kilometres (1,555 mi.) south of the geographical North Pole. Strangely, although this anthropological opus of nearly eight hundred pages was dedicated to the study of Inuit life, the opening chapter incorporated an article about the 'Pole of Relative Inaccessibility' published the year before in the *Geographical Review* of 1920. In it, Stefansson discussed the merits of Robert Peary's methods of reaching the North Pole. Stefansson had paid close attention to Peary's system, a method of travel using support teams with Inuit sledges and dogs. Stefansson admired Peary's highly organized approach to his expeditions – but only up to a point. Peary had relied on a military-style form of planning, which his critics have argued placed excessive demands on the resources of the Inuit families who themselves sacrificed a great deal for him to succeed. Only after time-consuming support to advance his position as far as possible towards the North Pole was his small party ready to travel as far as the pole. Using this system, Peary averaged about 19 kilometres (12 mi.) per day over the surface of the Arctic sea ice. With his base at Cape Columbia, only around 800 kilometres (500 mi.) from the pole, and able to traverse a relatively smooth section of the sea ice pack relatively free of obstructive pressure ridges, victory was his.[14]

Stefansson's opinion of Peary's achievement was that, with sufficient preparation and practice, anyone reasonably fit could also reach the geographical North Pole. Where the two men really differed is that Stefansson believed that Inuit, being people *of* and *from* the North, possessed the necessary skills and understanding to live at the North Pole in reasonable comfort, in stark contrast to transient visitors like Peary, whose next supply of food was the nearest depot.

Stefansson's discussion of Peary was accompanied by a dissection of four stages of polar exploration he perceived to characterize the North Pole. This classification scheme divided the Polar Basin into four quasi-evolutionary geographical zones, roughly tracing the material culture of European explorers as they had progressed northwards. These zones could be thought of as successive historical stages of technological mastery of the materials and skills of polar navigation. The first zone was primitive, corresponding to the distant murky past when Europeans had not yet learned to use the wind to propel their small boats, and so were at its mercy. The second zone represented the arrival of sailing technologies, enabling early modern explorers to use the wind to navigate up to the edge of the Arctic sea ice and record a succession of 'furthest north' points. Stefansson connected nine of these historic points reached by ships around the polar basin to give the best available indication of the ice edge. This line marked the southern border of the third zone, in which the Arctic sea ice was readily accessible to those who had borrowed dog and sledge technology from the Inuit. Stefansson then extended the point furthest north made by ships along the third zone's southern border in an arc of 800 kilometres, such that this third zone embraced all possible journeys equal in distance to Peary's 800-kilometre dash, including the geographical North Pole itself. For that reason, Stefansson called the third zone the 'area of comparative accessibility'.

Vilhjalmur
Stefansson's map
locates the North
Pole on the outer
contour as being much
easier to reach than
the Pole of Relative
Inaccessibility in the
centre of the Arctic
Ocean at 83° 50'.
In this inner zone,
Stefansson judged
that Inuit seal-hunting
skills would guarantee
the best chance of
survival, not Peary's
method of sledges,
dogs and depots.

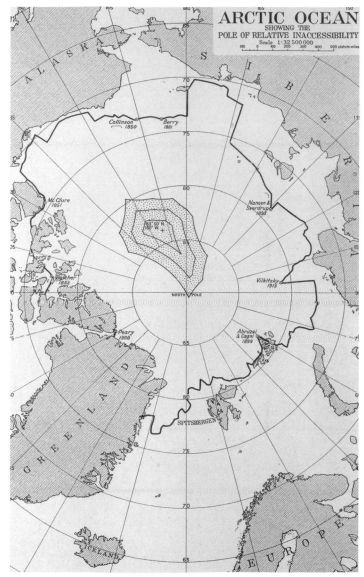

This, however, left an inner area in an irregularly shaped polygon on the ice pack in the middle of the Arctic Ocean. In contrast to the 'area of comparative accessibility' traversed by Peary, the shaded polygon represented Stefansson's fourth zone, an inner polar zone of 'comparative inaccessibility'. At the geometric centre of this polygon was the 'Pole of Inaccessibility', more than 640 kilometres (400 mi.) beyond the geographical North Pole.[15]

Peary had reached only the third of four stages or zones on the ladder of polar understanding. He had grasped the mechanics of Inuit methods of travel but had not embraced the wider philosophical understanding of Inuit mobility. Stefansson decided that the test of a fourth, higher stage of polar exploration was whether one could subsist at a new inner pole in the centre of the Arctic ice pack, beyond the geographical North Pole. No one had attempted to do so. The error of Peary and others was to fail to recognize that the Arctic Ocean was teeming with life on which he very likely could, and in Stefansson's view should, have subsisted. In his droll manner, Stefansson put it this way:

> to sum up, the arctic sea is lifeless except that it contains as much life to the cubic mile of water as any other sea [just as] the arctic land is lifeless except for millions of caribou and of foxes, tens of thousands of wolves and of musk oxen, thousands of polar bears, billions of insects and millions of birds.[16]

Stefansson named the contours of his oddly shaped zones 'isochrones', literally meaning places sharing the 'same time'. One recalls that in the framework inherited from the ancient Greeks, meridians or longitude were isochronic because celestial time is really just the movement of the heavens around the polar axis. However, Stefansson's isochrones were also 'hodological', meaning that they were relative to the contour of the ice edge, which itself was shaped by the vital material forces of the Earth.[17]

Stefansson's inner zone of comparative inaccessibility required the ability to forage by reading the hidden movements of sea life. This knowledge, tried and tested by centuries of hunting, meant that dedicated students of Inuit culture could inhabit the Pole of Inaccessibility. Stefansson believed that the lessons he had learned were ones that anyone could acquire in order to survive almost anywhere on the Arctic Ocean, being sustained by nature, rather than by subjugating it. Being able to pass beyond the zone of comparative accessibility traversed by Peary required not just mechanical hunting skills, but the understanding of a cosmology in which living with animals was a form of mutual dependence between humans, animals and spirits.[18] Thus in Stefansson's scheme, his fourth zone of polar exploration was really the Inuit zone – insightful and ironic, knowing that the sea ice held the key to sustaining the interconnection of life in the Polar Basin.

Stefansson was also saying that the imperial vision of Peary and those like him blinded them to the interiority of the 'living Arctic'.[19] Stefansson's inner isochronic zone remains closed to those lacking the necessary cultural understanding and skill. For all that Peary was greatly aided by his co-travellers, Matthew Henson and the four Inuit, he had travelled only as far as his experience, methods and beliefs permitted.

It is interesting to ask if Stefansson was constructing a utopia to challenge the dominant imperial values of his age, in ways that echo what Margaret Cavendish had done in *The Blazing World* some 250 years earlier. Humans can only access food in the zone of inaccessibility using Inuit methods and beliefs if the seals are willing to give themselves to the forager. Inuit recognize the spirits of animals as being persons, only of a non-human kind. Cavendish had been an avid and astute reader of natural philosophy, just as Stefansson was a serious student of the biology and nutrition of animals beneath the sea ice.[20] Stefansson explained to his readers that, in order for seals to inhabit Arctic sea ice, they must be able to surface to breathe by finding sufficiently thin ice to enable them to construct breathing holes, what the Inuit term *aglus*. He conjectured that if seals followed their breathing holes as the pack ice drifted across the Arctic Ocean from the Bering Strait area into the middle of the Polar Basin, then the area of comparative inaccessibility ought to be populated by seals. The relative ease of catching them meant they were the staple food of the polar marine world. It was as though Stefansson, somewhat tongue-in-cheek, had created the Pole of Inaccessibility to make the North Pole a colony, or at least an extension, of the Inuit world rich in sea life.

In this sense Stefansson's Pole of Inaccessibility might be considered a gateway to the living web of life beneath the ice where Sedna, the Inuit mother-spirit, looks after the souls of her children, the sea animals. When treated respectfully by humans, seals are said to offer themselves up to hunters who depend on them for nourishment; if angered by malevolent human misdeeds and a lack of respect, however, Sedna was known to withhold her animals' presence from hunters, leaving them to go hungry until they had confessed their transgressions. Clearly the rules of conduct of the area of Stefansson's inaccessible zone were going to be different to those emulating Herakles and the dash to the geographic North Pole. What worried Stefansson was the risk of encountering polar 'deserts', areas of sea ice too thick to allow seals to come up for air, and therefore lifeless and inaccessible. Analogous to the deserts of the temperate or tropical zones, these exceptional spaces couldn't be inhabited for sustained periods. They would need to be skirted or crossed by relying on Peary's 'pemmican-and-relay' method.[21]

## Poles of internationalism

Stefansson had been swimming against the military-industrial tide of mechanical power. His Pole of Inaccessibility was reached in 1926, but not by foraging. Instead, Roald Amundsen, world renowned for his South Pole triumph in 1912, planned an aerial approach, using for the first time an airship, an N-class semi-rigid dirigible. Teaming up with the airship's flamboyant Italian designer and Italian air force pilot, Umberto Nobile (1885–1978), and the American aviator Lincoln Ellsworth (1880–1951), the airship *Norge* and the sixteen-man crew left Ny-Ålesund in Spitsbergen on 11 May 1926 at 9.55 a.m. and arrived at the North Pole less than a day later at 1.25 a.m. Hours later, the *Norge* flew over the Pole of Inaccessibility, or as Amundsen called it, the 'Ice Pole', 'the hypothetical spot . . . hitherto unseen by man . . . at

Lat. 88°N, and Long. 157°E, the geographical centre of the Arctic ice mass'.[22]

The *Norge* represented the triumph of the Apollonian eye as the crew celebrated being *over* the North Pole. The expedition also meant different things to its members. Amundsen, Ellsworth and Nobile bore their nationalism with conviction, but Nobile's politics were fascist and he carried the blessing of Italy's dictator, Benito Mussolini. Aboard the dirigible, the anticipated celebration of international cooperation quickly became trumped by petty nationalism. When the three explorers released their national flags to fall to Earth over the geographical North Pole at the choreographed moment, Amundsen and Ellsworth instantly realized that they had been duped. Nobile had surreptitiously brought a flag far larger than theirs. The already tense relationship with Nobile grew much more strained. While the expedition served to show the potential for polar cooperation and internationalism, it also revealed that the narrower interests of nationalism could erupt from just beneath the surface.[23]

Umberto Nobile's expeditions brought together geopolitics, jealousy, intrigue and disaster. His dirigible (semi-rigid airship) *Italia* (1928) on its return from the North Pole crashed in the ice northeast of Svalbard.

The industrialization of the polar regions was unkind to Stefansson's vision of an Inuit-style Arctic Basin in which huskies were the ultimate explorers, using their strength and intelligence to pull sledges, and rewarded by feeding off fresh seal meat. Unfortunately for Stefansson the first dog to reach the Pole of Inaccessibility was a mere 25 centimetres (10 in.) tall and 5.5 kilograms (12 lb) in weight, and companion to a fascist. This small orphaned black-and-white fox terrier, Titina, was a homeless dog rescued by Nobile. Like her master, her bravery was applauded – it was said that she had won a stand-off with a polar bear by barking at it. The expedition's journalist, Antonio Quattrini, judged Titina to be 'a dog marked by destiny, a dog of [the] greatest character', in spite of her having 'taken a ferocious dislike' to his notebook, wanting to tear it to bits.[24] She was made into a celebrity, accompanied Nobile everywhere, and

The North Pole retained its sacred dimension even throughout the Cold War. In this photograph, the crew of the American USS *Skate* submarine are shown scattering the ashes of George Hubert Wilkins, who in 1930 had begun planning to take a submarine under the North Pole. The expedition failed, and the submarine was scuttled off the coast of Norway.

posed for photographs with Mussolini as well as President Coolidge and others.

As the definition of the Pole of Inaccessibility became understood as simply the point in the ocean most distant from land, with or without seals, it continued to be visited by expeditions carrying out research into sea ice, undertaking secretive military intelligence work and resupplying scientific drift stations. So, too, the development of submarines by the U.S. Navy was transforming Arctic oceanography. Seabed surveys were opening up a new world of undersea mountains, ridges and canyons. Geologists, seismologists and glaciologists began to investigate the strata of the Polar Basin, unlocking the history of its strata, the importance of former ice sheets and glaciers, and mapping the circular flow of ocean currents and sea ice across the Arctic Basin. Identifying mountain ranges and ridges also brought about the advent of submarine channels and routes, of growing importance from the 1950s onwards as the Cold War divided the Arctic into Soviet and American spheres of influence. This era of intensive military-industrial research led to an unprecedented investment in military infrastructure in the form of air-based radar stations, weather stations and ports, the imprint of which is very clearly visible in the infrastructure of the high Arctic today.

The Pole of Inaccessibility would in subsequent decades become a pole of diminished significance. Reflecting on the moment when the U.S. Navy submarine *Nautilus* had passed close by the Pole of Inaccessibility in 1958, the Master's comment was: 'but who cared? We were safe, warm, and comfortable in our home beneath the sea.'[25] This throwaway remark captured something interesting about the Pole of Inaccessibility. In spite of the growing geopolitical importance of the Arctic Basin as a strategic theatre in the Cold War, it had become at least for one sailor a mere curiosity, holding little in the way of prestige or significance.

## The revival of the geographical North Pole

The North Pole became more accessible from the mid-1920s onwards largely because polar approaches became redefined by the use of new technologies. How one approached the North Pole became as important as whether one had reached its precise coordinates. The relationship between technology and logistics, central to the expansion of trade and empire in the 1930s, became a defining feature of the twentieth-century polar microworld. Techniques of handling dog teams, paddling kayaks and landing aircraft on ice became emblematic of the specialized skills of polar research. Possessing the skills to be self-contained and autonomous in polar fieldwork acquired new prestige as measures of masculine polar character and leadership. Risks were to be calculated and managed. The provisioning of food, the incorporation of indigenous clothing for its unique properties of humidity and temperature control, and experiments with new kinds of locomotion and measurement were all part of the growing emphasis on expertise in polar fieldwork.[26]

Scientific studies of the atmosphere and ocean were developing models of the Arctic as a three-dimensional volume, no longer just a surface plane. Incursions through air and seawater by aeroplanes and submarines redefined the experience of polar travel that experiments with hot-air balloons had long hinted at. Explorers and journalists were challenged to conceptualize new visual narratives for depicting these feats, so that readers could picture the new polar topography with kilometre-length ice cores, high-atmosphere wind systems, transpolar radio communications, and seabed ridges and canyons.[27] This was a polar world in which knowledge and access were becoming governed by advanced technological systems.

If during the interwar period the geographical North Pole had become an object of nostalgia, it made a remarkable comeback in the geopolitical struggles of the Second World War and the Cold War. In a world regionally defined by geopolitical divides and technological power, a polar vision offered a visual and spatial way of framing the political world order, and a spatial

A tent from Ivan Papanin's Russian Drift station, North Pole-1, inspired by Nansen's *Fram* expedition (1893–96), drifted from near the North Pole for nine months on the pack ice a distance of 2,850 km before being picked up by icebreakers. The scientific party was much celebrated and awarded the title of 'Hero of the Soviet Union'.

vocabulary for reflecting on the fate of empires and superpowers. Unsurprisingly, maps drawn on a polar projection were popular among military planners and analysts. Elmer Plischke, a u.s. Navy historian and politics professor, enthused about the practical 'ease with which rectilinear air routes between all points of the globe can be observed'.[28] Polar maps also acquired a new realism in the geographical imagination. Journalists too could use the projection's Apollonian view from above as though it were the bird's-eye view of a pilot looking down over the top of the world. That Plischke attributed the popularity of the polar projection to it being 'recently designed' was a sign of the polar map's fit with technological modernity.

The seamless integration of the geographical North Pole into Arctic airspace could be made to illustrate a world order marked by either national divisions or international cooperation: a new 'global aerialism'. Following the Battle of Guadalcanal in February 1943, Allied forces moved on to the offensive in the Pacific War against Japan. That summer *Life* magazine ran a feature warning against the isolationist politics that 'lulled us into a false sense

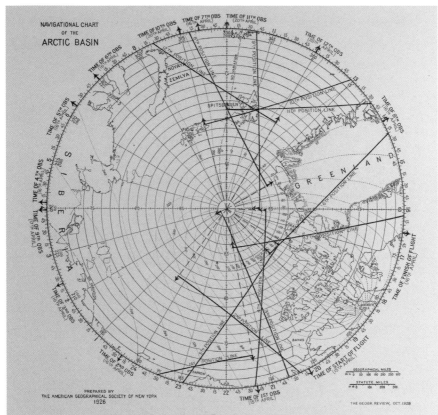

NAVIGATIONAL CHART OF THE ARCTIC BASIN

PREPARED BY
THE AMERICAN GEOGRAPHICAL SOCIETY OF NEW YORK
1926

THE GEOGR. REVIEW, OCT. 1928

FIG. 17

| OBSERVATION | DATE AND GREENWICH APPARENT CIVIL TIME | | OBSERVED ALTITUDE | SUN'S DECLINATION | POLAR DISTANCE OF POSITION LINE ON SUN'S MERIDIAN | | DEDUCED GROUND SPEED IN STATUTE MILES PER HOUR |
|---|---|---|---|---|---|---|---|
| Flight begun | 15/4 | 20$^H$ 10$^M$ | | | | | |
| No. 1 | 15/4 | 21$^H$ 20$^M$ | 26° 30' | 9° 54' | +16° | 36' | 111 |
| No. 2 | 16/4 | 0$^H$ 22$^M$ | 19° 11' | 9° 57' | + 9° | 14' | 117 |
| No. 3 | 16/4 | 2$^H$ 22½$^M$ | 10° 48' | 10° 00' | + 0° | 48' | 122 |
| No. 4 | 16/4 | 4$^H$ 07½$^M$ | 5° 32' | 10° 00' | − 4° | 28' | 106 |
| No. 5 | 16/4 | 6$^H$ 16½$^M$ | 2° 50' | 10° 02' | − 7° | 12' | 88 |
| No. 6 | 16/4 | 7$^H$ 0$^M$ | 3° 00' | 10° 03' | − 7° | 03' | 87 |
| No. 7 | 16/4 | 9$^H$ 05$^M$ | 5° 35' | 10° 05' | − 4° | 30' | |
| No. 8 | 16/4 | 14$^H$ 04$^M$ | 15° 32' | 10° 10' | + 5° | 22' | 129 |
| Flight ended | 16/4 | 16$^H$ 30$^M$ | | | | | 197 |
| No. 9 | 17/4 | 3$^H$ 58$^M$ | 6° 37'? | 10° 21' | − 3' | −44' | |
| No. 10 | 20/4 | 9$^H$ 07$^M$ | 21° 29' | 11° 29' | + 9° | −38' | |
| No. 11 | 20/4 | 10$^H$ 0$^M$ | 22° 30' | 11° 30' | +11° | −00' | |
| No. 12 | 20/4 | 12$^H$ 0$^M$ | 22° 35' | 11° 32' | +11° | −03' | |

Hubert Wilkins's 'Navigational chart of the Arctic Basin', *Geographical Review* (1928). With the advent of 20th-century aviation, polar charts gave the vision of early cosmographers a new technological realism, enabling North Pole bearings to be drawn as straight lines, as though looking down on the Polar Basin from a high altitude.

of security before Pearl Harbor'. 'The startling truth' was that, 'today, because of the plane, no spot in the world is more than 60 hours' flying time away from your local airport'. *Life* claimed that the 'North Polar projection of the world' being taught to schoolchildren, 'if they are honest maps, will clearly tell us we can no longer cling to the old-fashioned "two-hemisphere" idea of geography'. This 'air-map of the world . . . shows us the world as it really is – a world without fences . . . in which nations once-remote are now clustered together in one global community'. Tomorrow's maps will instead show 'planes of peace and commerce'.[29]

The aspiration to celebrate the prospect of overcoming geo-political strife with images of circumpolar unity was utopian. *Life* magazine's anticipation of an Allied victory leading to a 'Free World' around a pole of peace was, like all utopias, premature. Polar maps joining East and West in the Cold War heightened the spatial sense of political tension rather than smoothing it over. They projected a space that could be simultaneously read as smooth and unified, while framing it as divided in two halves, usually placing North America on the left and Asia on the right. Military strategic planning made extensive use of polar views to map the reach of American and Soviet long-range bombers, over-the-horizon radars, nuclear-powered submarines and ballistic missiles. Transpolar commercial air routes also became a reality in this context. The Soviet Union began running scheduled flights twice a week to Cuba in the mid-1960s using a turboprop airliner on the transpolar Moscow–Havana route. Unscheduled stopovers occasionally became a necessity, as when one Russian flight was permitted to refuel at John F. Kennedy airport in New York, before going on its way to Cuba.[30]

The renewed popularity of the polar projection during the Second World War, in promoting a vision of the world's peoples gathering together in peace at the pole, invoked a cartographic tradition with its roots in sixteenth-century cosmography and politics. Even if the principal purpose of cosmography – to study the harmonies between the Earth and the heavens – had been discarded, the Apollonian vision of flying over the globe persisted even as ways of envisioning the globe changed many times. The techniques developed by Dürer and Apian of drawing maps with an external eye in line with the polar axis had a lasting impact on popular and political cartography. The use of polar cosmography to map the eastern and western hemispheres, testifying to the global scale of Charles v's universal power, fascinated and inspired polar explorers like Peary and Nordenskiöld in a period of high empire nearly four hundred years later. So, too, the system of defining time and place in terms of zones and measures of longitude was a lasting legacy of cosmography in shaping new institutions and technologies of global standards. These ways of envisioning time and space were invoked time and again, albeit in new ways, by later generations and have had a lasting impact on what is often loosely described as 'Western' and 'global'.

## The North Pole's new narratives

The geographical North Pole, by virtue of being a point of projection, a source of longitude yet having no time or longitude of its own, and set in a location difficult to scrutinize, has long tempted rulers, religions and occultists as a place well suited to host their utopian frameworks. Satire and other forms of unauthorized writing can reach places where others cannot. Authors writing from the margins often claim that the dice are loaded in favour of authoritarian, nationalist or corporate visions. That may help explain why poles remain lively conductors for New Age movements, theories of past and impending cataclysms, and astrological visions of a unified global village.

The poles still retain a power for genuine philosophical reflection. One reason is that since the time of Nansen's drawing of the Polar Basin, the sciences have produced an astonishingly rich set of accounts revealing how the dynamics of the cryosphere, marine and terrestrial environments of the polar regions play a critical role in sustaining the planet's global-scale systems – even if the marine environment around the North Pole itself is not yet well documented. Harnessing evidence from polar science to inform policy debates at a time of rapid climate change and to communicate to wider audiences is a task of central importance.[31]

The Arctic journeys of the most thoughtful twenty-first-century explorers – particularly indigenous peoples, scientists, artists and writers – have retained a philosophical and ethical engagement with the Arctic's capacity to reflect the mutual dependence that ties our fate to the living world.[32] Vilhjalmur Stefansson frequently observed that he reached his very large audiences best by using empiricism and analysis to puncture false myths, and to use the facts of science to communicate with allegory. Indigenous peoples of the Arctic have also recognized that they can communicate the importance of safeguarding polar ecosystems to the rest of the globe if they have philosophical as well as practical consequences for listeners or readers wanting to understand what it means to inhabit this planet. Those meanings are often best understood by bringing together allegories, narratives and observations. When audiences see or read reports about polar bears spending more time on shore in the face of melting summer sea ice, the migration of fish stocks

Snowshoes, originally an indigenous technology of travel for trapping or foraging, were used in the North Pole marathon until 2007. The modern aluminium designs were ruled out to avoid hip flexor injuries and to protect the tent floors from tearing.

The annual North Pole Marathon, created in 2002 by Richard Donovan and supported by Russian field operators, is a contemporary festive ritual of polar circumnavigation on foot. Competitors join hands in a show of cooperative transnationalism. No visas are required for the North Pole.

towards the North Pole as the temperature of the oceans rise, or the dilemmas of people seeking refuge where the melting tundra around their settlements has begun to collapse, narrative and fact work hand in hand, and pose important questions to people in very different settings. That is one of the important ways in which sense is made of the diverse geographies of the globe.

The North Pole Marathon is a very apt symbol of the Pole's global cultural politics. The idea that, once upon a time, began as a heroic and life-consuming run to warn the Athenians of an impending invasion, is today staged at the North Pole. The course, comprising a number of laps, is really a festival of hardy runners and dreamers coming together in a spirit of international cooperation and mutual support. The not insignificant infra-structure – a runway and camp – are provided with the support of Russian logistics experts. And yet the staging of the marathon by those who seek an extraordinary setting remains a harbinger of danger. The pack ice has always been drifting beneath the feet of polar approachers, marathon runners included, but climate change is altering the sea ice, warming the atmosphere, acidify-ing the ocean and changing the polar marine ecosystem.

Stefansson's term 'Pole of Inaccessibility' would be an apt description of the gendering of polar fieldwork, both Arctic and Antarctic, in the twentieth century. Pioneering women like Marie Stopes (1880–1958), who applied to take part in Robert Falcon Scott's South Pole expedition, were turned down point blank.[33]

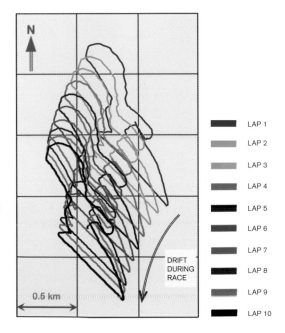

During the time taken to complete the ten laps of the North Pole Marathon (2007), the sea current driving the sea ice moved the course approximately a kilometre to the south.

Paralympic neuroscientist, physician and athlete Dr William Tan completed the North Pole Marathon course in 21 hours and ten minutes using the graded surface of the aircraft runway, enduring -25°c temperatures. In spite of holding multiple Paralympic world records, he prioritizes fundraising for humanitarian projects for a global vision based on a greater common good.

The married status of some women allowed them to accompany their husbands on polar travels. However, the British Antarctic Survey, for example, permitted women scientists to overwinter at one of its research stations on the Antarctic continent only as recently as 1994.[34] The Arctic, having a long history of colonized peoples and frontiers, has had a much wider range of roles for women, many of them relatively independent, both among settlers and their indigenous hosts, who were of course women as well as men.[35]

With polar approaches traditionally defined by men raised with muscular, imperial and competitive values, women explorers have brought their own distinctive narratives, though generally reworking the existing framework of competition, muscle and 'firsts'. Rather than abandoning the North Pole conquests as a busted flush, women have used markers of gender, ethnicity and priority to rewrite the record books. Ann Bancroft (*b.* 1955) became the first woman to reach the North Pole by dog sled in 1986 as a member of Will Steger's five-person team.[36] That women could envision a meritocratic approach to selecting partici-pants was demonstrated when film financier Caroline Hamilton hatched a plan for an all-women team to ski pulling sledges to the North Pole. She advertised trials for twenty women to form an international polar relay team to make the 670-kilometre (416-mi.) journey to the North Pole – with no polar experience necessary.[37] This challenged traditional readings of women's roles in unexpected ways, though press coverage has often insisted on highlighting gender when many women simply want to be known for their accomplishments. Members of Hamilton's relay expedition were reported not only as adventurers, but as mothers who had left children at home, as female amateurs and as a mother-daughter team.

In 2007 retired nurse Barbara Hillary (*b.* 1931) skied from the 89th parallel to the North Pole at age 75, after surviving lung and breast cancer. An African American raised in poverty in Harlem, she had overcome many obstacles throughout her life, and her lack of access to skiing was just one of them. By organizing coaching sessions at the gym, taking skiing lessons in Canada and hiring a guide, she made it to the North Pole. Four years later, she turned her sights to the Antarctic and became the first African American, male or female, to reach both poles.[38] Her achieve-ments bring into focus the intersection of issues of race, gender and class. They amplify, for example, the recognition eventually given to Matthew Henson's very remarkable achievements as a polar traveller in his own right.[39] One way in which Hillary has

authorized the symbolism of the North Pole (though by no means the only reading) is as a source of hope to others battling cancer who may find courage to take on their own challenges.

In the present era of the Arctic experiencing rapid climate change, most polar approaches have of necessity focused on the instability of the diminishing sea ice. In one of the most thoughtful expeditions, the Dutch adventurer and journalist Bernice Notenboom (*b*. 1962) turned the concept of the polar approach on its head. Instead of filming her approach, she and filmmaker Sarah Robertson chose to film her journey skiing away from the North Pole.[40] Thus she emphasized the relationship with the North Pole as one of moving away and receding, rather than the centuries-old tradition of moving towards and beyond. No historical painter during the Renaissance, or since, has portrayed Herakles leaving after defeating Antaeus. Did Herakles turn his back on him or attend to his remains before resuming his pursuit of the golden apples? A number of Arctic adventurers have

Ann Daniels and Caroline Hamilton making the first British all-women ski expedition from land to the North Pole, in 2002, shown here following the line of an ice ridge to the pole.

observed that the reduction in Arctic sea ice is making polar approaches much more difficult. Notenboom, rather than seeking media coverage for yet another approach, instead drew attention to the likelihood that she might be the last person to ski away from the North Pole, such were the deteriorating ice conditions. Thus in her film *Sea Blind*, made with Robertson, viewers have an opportunity to reflect on the loss of sea ice, not just as the end of a pastime, but a loss to mourn, and yet still a call to action. Such expeditions go further than spreading the message that the polar regions play a critical role in sustaining this planet. They invite us to reflect on Antaeus' secret, his history, now uncovered, unable to renew his strength, incapable of repelling hubris. If in our Arctic narratives we persist in identifying with Herakles in the twenty-first century, is it possible or desirable for us to mourn Antaeus in the way that his mother Gaia deserves reverence?

Ann Daniels following a centuries-old tradition dating back to Scoresby and Parry, manoeuvres her amphibious sledge across an open water lead en route to the North Pole.

What are we to make of his power to fashion a temple for his father, Poseidon, from the skulls of those he vanquished? Perhaps then it can be an invocation to reflect on the way that certain kinds of stories through the age have, for better or worse, shaped relationships between fathers and sons.

The tragedy is that while we recognize that our fate is bound up with the engine of late industrial capitalism, the world continues to identify with the hubris of Herakles. The Arctic is part of this tragedy, with toxic contaminants polluting land, sea and ocean. The Barents Sea off the coast of Norway and Russia is reportedly the most radioactive in the world, largely as a result of nuclear atmospheric tests carried out during the Cold War, emissions from reprocessing plants and the accident at Chernobyl.[41] To the north and east of the Greenland and Barents Seas, high concentrations of old plastic arrive from the Atlantic Ocean by thermohaline ocean circulation, which acts as a 'plastic conveyor belt' from distant sources.[42] In light of the fragile networks of circumpolar cooperation in science, education and governance, what would it look like instead to view the labour

Bernice Notenboom practises environmental philosophy as an experienced polar traveller. Her approach to philosophical reflection on questions of cryopolitics and global environmental concern takes place by journeying through the changing materiality of the polar world.

Turning on its head the centuries-old tradition of polar approaching, in 2014 Bernice Notenboom set out to ski away from the North Pole to see whether the sea ice conditions would allow her to reach land. In this photography she takes stock of multi-year sea ice: majestic and sublime, yet precarious and endangered.

of Herakles from the perspective of the damaged Antaeus? What if we were to identify with King Euresthus who, having sent Herakles to steal the golden apples of the Hesperides, recanted on his wish for immortality and restored the golden apples to their rightful owner?[43]

It is perhaps fitting to finish the story of poles and polarity through a reading of Antaeus closer to the home of the author. In 'Antaeus' the myth is reworked from the inside out by Seamus Heaney, the Nobel Laureate poet, writing about the history of the Northern Ireland Troubles. There will be different readings of the poem, but a more ecological reading than perhaps was intended is to reflect on how one stays grounded in order for a society to remain unyielding to the destructive and unquenchable appetite of a colonizing world:

> Let each new hero come
> Seeking the golden apples and Atlas:
> He must wrestle with me before he pass
> Into that realm of fame.
>
> Among sky-born and royal.
> He may well throw me and renew my birth
> But let him not plan, lifting me off the earth,
> My elevation, my fall.[44]

# REFERENCES

## Preface

1 John Barrow, *A Chronological History of Voyages into the Arctic Regions* (London, 1818).
2 J. Lennart Berggren and Alexander Jones, ed., *Ptolemy's Geography: An Annotated Translation of the Theoretical Chapters* (Princeton, NJ, 2000).
3 Denis Cosgrove, *Apollo's Eye: A Cartographic Genealogy of the Earth in the Western Imagination* (Baltimore, MD, 2001).
4 Claudio Aporta and Eric Higgs, 'Satellite Culture: Global Positioning Systems, Inuit Wayfinding, and the Need for a New Account of Technology', *Current Anthropology*, XLVI (2005), pp. 729–53.
5 Richard Dunn and Rebekah Higgitt, *Ships, Clocks, and Stars: The Quest for Longitude* (London, 2014).

## 1 The Upward Gaze

1 Claudio Aporta, 'The Trail as Home: Inuit and their pan-Arctic Network of Routes', *Human Ecology*, XXXVII (2009), pp. 131–46.
2 Ibid.
3 Claudio Aporta and Eric Higgs, 'Satellite Culture: Global Positioning Systems, Inuit Wayfinding, and the Need for a New Account of Technology', *Current Anthropology*, XLVI (2005), pp. 729–53.
4 Aporta, 'The Trail as Home'.
5 David Damas, ed., *Handbook of the North American Indians*, vol. V: *Arctic* (Washington, DC, 1985).
6 Dan Laursen, 'Obituary: Eigil Greve Knuth, 1903–1996', *Arctic*, XLIX/4 (1996), pp. 401–3.
7 John MacDonald, *The Arctic Sky: Inuit Astronomy, Star Lore, and Legend* (Toronto, 1998).

8  Ibid., p. 60.

9  Ibid., pp. 171–2.

10  Ibid., p. 172.

11  Ibid., p. 170.

12  Ibid., p. 167. MacDonald cites a personal communication from Duncan Pryde, 20 July 1995.

13  Robert Peary, *Northward over the 'Great Ice': A Narrative of Life and Work along the Shores and upon the Interior Ice-cap of Northern Greenland in the Years 1886 and 1891–97*, 2 vols (London, 1898), I, p. 494.

14  Michael Robinson, *The Coldest Crucible: Arctic Exploration and American Culture* (Chicago, IL, 2006).

15  Ibid., pp. 98–110.

16  Karen Morin, *Civic Discipline: Geography in America, 1860–1890* (Farnham, 2011), pp. 29–53.

17  MacDonald, *Arctic Sky,* pp. 55–6 for the Kingulliq legend, and pp. 230–31 for the Iliarjugaarjuk legend.

18  Denis Cosgrove, *Apollo's Eye: A Cartographic Genealogy of the Earth in the Western Imagination* (Baltimore, MD, 2001), pp. 30, 43; James Romm, *The Edges of the Earth in Ancient Thought* (Princeton, NJ, 1992), pp. 62–9.

19  Robert E. Peary, *The North Pole: Its Discovery in 1909 under the Auspices of the Peary Arctic Club* (New York, 1910), p. 267.

20  Ibid., p. 266.

21  Ibid. The classic study of rituals of possession is Patricia Seed, *Ceremonies of Possession in Europe's Conquest of the New World, 1492–1640* (Cambridge, 1995).

22  Peary, *Northward over the 'Great Ice'*, p. 494.

23  Tim Ingold, *The Perception of the Environment: Essays on Livelihood, Dwelling and Skill* (Abingdon, 2001), pp. 219–42.

24  'Precession' is the term used to describe the change in the orientation of the axis of a spinning body such as the Earth. This is what causes the position of the pole star (and the other bodies) to shift over time. In ancient Greek astronomy, which assumed that the Earth was fixed, precession was visible in the changing position of the stars in the celestial sphere. See www.wikipedia.org.

25  For 'Polaris' and 'celestial pole', see the respective entries at www.wikipedia.org.

26  Peary, *North Pole*, p. 266.

27  'The Incredible Story of Inuit Fathered by a U.S. Explorer and his Aide who are Making Their Way in a Globalized World', www.dailymail.co.uk, updated 12 September 2011. Matthew Henson is remembered and celebrated by his descendants among the Inughuit

of northwest Greenland. In the United States Henson is also widely used as a historical role model in many exploration stories for young people.

28 Thomas Ingham, 'Exploring a Heritage: Tracing Victorian Thoughts on Ancient Exploration', BA diss., Department of Classics, University of Cambridge, 2017.

29 Hipparchus, *Commentary on the Phaenomena of Aratus and Eudoxus*, ed. and trans. C. Manitius (Leipzig, 1894), 1.4.1; see also Gerald J. Toomer, 'Hipparchus', in *Oxford Classical Dictionary*, 3rd edn (Oxford, 1986).

30 Thomas Gladwyn, *East is a Big Bird: Navigation and Logic on Puluwat Atoll* (Cambridge, MA, 1970); Ward H. Goodenough, ed., 'Prehistoric Settlement of the Pacific', *Transactions of the American Philosophical Society*, LXXXVI (1996), pp. 1–10.

31 Liba Taub, *Ptolemy's Universe: The Natural Philosophical and Ethical Foundations of Ptolemy's Astronomy* (Chicago, IL, 1993), p. 141.

32 Ibid., pp. 143–4.

33 Ibid., p. 136.

34 Ibid., pp. 137–8.

35 Ibid., p. 144.

36 J. Lennart Berggren and Alexander Jones, ed., *Ptolemy's Geography: An Annotated Translation of the Theoretical Chapters* (Princeton, NJ, 2000), p. 58.

37 Ibid., p. 28.

38 Ibid., p. 13.

39 Romm, *The Edges of the Earth*, p. 61.

40 Ibid., p. 62.

## 2 Holding the North Pole

1 George Kish, 'Peter Apian', in *Dictionary of Scientific Biography*, revd edn (New York, 2008), I, pp. 178–9; Siegmund Günther, *Peter und Philipp Apian, zwei Deutsche Mathematiker u. Kartographen*, Abhandlungen der Königlich böhmischen Gesellschaft der Wissenschaften, 6th ser., XI (Prague, 1882).

2 Johannes Keunig, 'The History of Geographical Map Projections until 1600', *Imago Mundi*, XII (1955), p. 8; John P. Snyder, *Flattening the Earth: Two Thousand Years of Map Projections* (Chicago, IL, 1993), pp. 1–24.

3 Jerry Brotton, 'Terrestrial Globalism: Mapping the Globe in Early Modern Europe', in *Mappings*, ed. Denis E. Cosgrove (London, 1999), pp. 71–89.

4 Bernard Goldstein, 'The Survival of Arabic Astronomy in Hebrew', *Journal for the History of Arabic Science*, III (1979), pp. 31–9; Owen

Gingerich, 'Islamic Astronomy', *Scientific American*, CCLIV (1986), pp. 74–83; Jerry Brotton, *The Renaissance: A Very Short Introduction* (Oxford, 2006).

5  Peter Apian, *Astronomicum Caesareum* (Ingoldstadt, 1540).

6  Peter Apian, *Cosmographicus liber* (Landshut, 1524).

7  Claudius Ptolemy, *In hoc opere continentur Geographiae*, trans. Johannes Werner (Nuremberg, 1514).

8  Adele Wörz, 'The Visualization of Perspective Systems and Iconology in Dürer's Cartographic Works: An In-depth Analysis Using Multiple Methodological Approaches', PhD diss., University of Oregon, 2006.

9  Oxford Museum of the History of Science, 'The Astrolabe: An Online Resource', www.mhs.ox.ac.uk, 1 October 2017.

10  Jim A. Bennett, *The Divided Circle: A History of Instruments for Astronomy, Navigation, and Surveying* (Oxford, 1987), pp. 33–40, 77–9.

11  Ptolemy, *In hoc opere continentur Geographiae*.

12  Nicholas Crane, *Mercator: The Man who Mapped the Planet* (London, 2002); William H. Sherman, *John Dee: The Politics of Reading and Writing in the English Renaissance* (Amherst, MA, 1995).

13  Owen Gingerich, 'Astronomical Paper Instruments with Moving Parts', in *Making Instruments Count*, ed. R.G.W. Anderson, J. A. Bennett and W. F. Ryan (London, 1993), pp. 63–74.

14  Apian, *Cosmographicus liber*; Peter Apian, *Horoscopion Apiani* (Ingolstadt, 1533).

15  Gingerich, 'Astronomical Paper Instruments'.

16  Ibid.

17  Jim A. Bennett, 'Sundials and the Rise and Decline of Cosmography in the Long Sixteenth Century', *Bulletin of the Scientific Instrument Society*, CI (2009), pp. 5–6.

18  Apian, *Cosmographicus liber*, frontispiece.

19  Apian's iconography largely affirms the classic argument about relations of power and the construction of the Orient in the imagination of the West as feminine and submissive; see Edward Said, *Orientalism* (London, 1978).

### 3 The Multiplication of Poles

1  Stephen Pumfrey, *Latitude and the Magnetic Earth* (Cambridge, 2002), pp. 11–23. My understanding of Gilbert's magnetical philosophy is indebted to Pumfrey's scholarship.

2  Ibid., p. 114.

3  Ibid.

4 Ibid.

5 Ibid., p. 113.

6 Ibid., pp. 14–16.

7 Ibid., p. 110.

8 Ibid., pp. 89, 111, 123–5.

9 Ibid., pp. 42–53.

10 Ibid., pp. 98–108.

11 Ibid., pp. 123–6.

12 Art Jonkers, 'North by Northwest: Seafaring, Science, and the Earth's Magnetic Field (1600–1800)', PhD diss., Vrije Universiteit Amsterdam, 2000.

13 Pumfrey, *Latitude*, pp. 159–72.

14 Ibid., p. 106.

15 Ibid., pp. 42–53.

16 Stephen Pumfrey, 'The *Selenographia* of William Gilbert: His Pre-telescopic Map of the Moon and his Discovery of Lunar Libration', *Journal for the History of Astronomy*, XLII/2 (2011), pp. 193–203.

17 William Gilbert, *De Magnete* (London, 1600), book 1 chap. 2 (p. 9; my translation).

18 Pumfrey, *Latitude*, p. 222.

19 Ibid., pp. 54–9.

20 Ibid., p. 115.

21 John Cawood, 'The Magnetic Crusade: Science and Politics in Early Victorian Britain', *Isis*, LXX (1979), pp. 492–518; Trevor Levere, *Science and the Canadian Arctic: A Century of Exploration, 1818–1918* (Cambridge, 1993), pp. 142–8.

22 Thomas James, *The Strange and Dangerous Voyage of Captaine Thomas James in his Intended Discovery of the Northwest Passage into the South Sea . . .* (London, 1633), sigs Q–Qv.

23 Ibid., sigs Q–Qv.

24 Ian S. MacLaren, 'Booking a Northwest Passage: Thomas James and *The Strange and Dangerous Voyage* (1633)', in *The Quest for the Northwest Passage*, ed. Frédéric Regard (London, 2013), pp. 89–102.

25 Joseph Moxon, *A Brief Discourse of a Passage by the North-pole to Japan, China, etc. pleaded by Three Experiments . . .* (London, 1674).

26 Edmond Halley, 'A Theory of the Variation of the Magnetic Compass', *Philosophical Transactions of the Royal Society*, XIII (1683), p. 219.

27 Ibid., pp. 208–21.

### 4 Polar Voyaging

1 William E. Parry, 'Expedition to the North Pole', *The Times*, 17 April 1827, p. 3.

2 William Scoresby Jr, 'On the Polar Ice', *Memoirs of the Wernerian Natural History Society*, II (1815), pp. 261–338.

3 Daines Barrington, *The Possibility of Reaching the North Pole Discussed* (London, 1775).

4 [John Barrow], 'Burney – Behring's Strait and the Polar Basin', *Quarterly Review*, XVIII (January 1818), pp. 431–58. The quotation about 'frenzied speculation' originates from Scoresby, 'On the Polar Ice', p. 328.

5 W. Scoresby Jr, 'On the Probability of Reaching the North Pole', *Edinburgh New Philosophical Journal*, V (1828), pp. 22–42.

6 J. Mirsky, *To the Arctic! The Story of Northern Exploration from Earliest Times* (London, 1944).

7 John McCannon, *A History of the Arctic: Nature, Exploration and Exploitation* (London, 2012); Richard Vaughan, *The Arctic: A History* (Stroud, 2007).

8 Mirsky, *To the Arctic*, p. 222.

9 Michael Robinson, *The Coldest Crucible: Arctic Exploration and American Culture* (Chicago, IL, 2006).

10 Ann Savours, '"A Very Interesting Point in Geography": The 1773 Phipps Expedition towards the North Pole', *Arctic*, XXXVII (1984), pp. 402–28.

11 Adriana Craciun, *Writing Arctic Disaster: Authorship and Exploration* (Cambridge, 2016), p. 28.

12 For the Longitude Act reforms of 1776, see 16 Geo. III c. 6; for the Longitude Act reforms of 1818, see 58 Geo. III c. 20. These can be consulted at the Board of Longitude Digital Archive at http://cudl.lib.cam.ac.uk, accessed 13 December 2017.

13 Barrington, *The Possibility of Approaching the North Pole Discussed*, pp. 115, 133.

14 Michael Robinson, 'Reconsidering the Theory of the Open Polar Sea', in *Extremes: Oceanography's Adventures at the Poles*, ed. Keith R. Benson and Helen M. Rozwadowski (Sagamore Beach, MA, 2007), pp. 15–29.

15 Act Geo. III, 1818, available at the Board of Longitude Digital Archive, http://cudl.lib.cam.ac.uk.

16 Sophie Waring, 'Thomas Young, the Board of Longitude and the Age of Reform', PhD diss., University of Cambridge, 2014; David Phillip Miller, 'The Royal Society of London, 1800–1855: A Study in the Cultural Politics of Scientific Organization', PhD diss., University of Pennsylvania, 1981.

17 Helmut Müller-Sievers, *The Science of Literature: Essays on an Incalculable Difference* (Berlin, 2015).

18 Scoresby, 'Polar Ice', p. 222; Tom Stamp and Cordelia Stamp, *William Scoresby Jr: Arctic Scientist* (Whitby, 1976), p. 52.

19 William E. Parry, *Narrative of an Attempt to Reach the North Pole* (London, 1828), pp. xi–xiii.

20 Scoresby, 'On the Probability of Reaching the North Pole', p. 25.

21 Parry, 'Expedition to the North Pole', p. 3. See also report in *Liverpool Mercury*, 5 January 1827, citing the use of the baidar from Barrow's notice in the *Quarterly Review*.

22 Robinson, *Coldest Crucible*, p. 1.

23 Mirsky, *To the Arctic*, p. 297.

24 William E. Parry, *Narrative of a Second Voyage in Search of a Northwest Passage* (London, 1824), p. 222.

25 Parry, 'Expedition to the North Pole'.

26 Parry, *Narrative of an Attempt to Reach the North Pole*, pp. 53–5, 92.

27 John Barrow, 'Narrative of an Attempt to Reach the North Pole' [Review of Parry], *Quarterly Review*, XXXVII (1828), pp. 523–39.

28 Janice Cavell, *Tracing the Connected Narrative: Arctic Exploration in British Print Culture, 1818–1860* (Toronto, 2008). For example, Cavell provides a valuable discussion of the phenomenal publishing success of Francis McClintock's *The Voyage of the Fox: A Narrative of the Discovery of the Fate of Sir John Franklin and his Companions* (London, 1859).

29 For the publishing context and strategy of Peary, see Robinson, *Coldest Crucible*, pp. 124–57.

30 Kirsten Thisted, 'Voicing the Arctic: Knud Rasmussen and the Ambivalence of Cultural Translation', in *Arctic Discourses*, ed. Anka Ryall, Johan Schimanski and Henning H. Wærp (Newcastle, 2010), pp. 59–81.

31 William Beechey, *A Voyage of Discovery Towards the North Pole . . . in 1818* (London, 1843). A classic analysis of the aesthetics of Arctic landscapes in the Victorian era is I. S. MacLaren, 'The Aesthetic Map of the North, 1845–1859', *Arctic*, XXXVIII (1985), pp. 89–103.

32 Barrow in fact attacks his critics by accusing them of being interested in the pursuit of treasure or acting out of self-interest alone; see Barrow, 'Narrative of an Attempt to Reach the North Pole', p. 534.

33 Ibid., p. 537.

34 Craciun, *Writing Arctic Disaster*, p. 28; Cavell, *Tracing the Connected Narrative*.

35 Russell A. Potter, *Arctic Spectacles: The Frozen North in Visual Culture* (Seattle, WA, 2007), pp. 41–6.

36 Frederick Beechey, *A Voyage of Discovery towards the North Pole: Performed in His Majesty's Ships Dorothea and Trent . . .* (London, 1843).

37 Huw Lewis-Jones, *Imagining the Arctic: Heroism, Spectacle, and Polar Exploration* (London, 2017)

38 Cavell, *Tracing the Connected Narrative*, pp. 151–2. See, for example, William Edward Parry, *Three Voyages for the Discovery of a Northwest Passage . . . and Narrative of an Attempt to Reach the North Pole* (London, 1831).


**5 Polar Edens**

1 Joscelyn Godwin, *Arktos: The Polar Myth in Science, Symbolism, and Nazi Survival* (London, 1996), pp. 19–24.

2 Ibid., p. 23.

3 Ibid., p. 20.

4 Richard Drayton, *Nature's Government: Science, Imperial Britain, and the 'Improvement' of the World* (Cambridge, 2000).

5 Godwin, *Arktos*, pp. 19–24.

6 Edgar Allan Poe, *The Narrative of Arthur Gordon Pym, of Nantucket, North America: Comprising the Details of a Mutiny, Famine, and Shipwreck, during a Voyage to the South Seas* (London, 1838); Jules Verne, *At the North Pole; or, The Adventures of Captain Hatteras* (Philadelphia, PA, 1874); Jules Verne, *The Purchase of the North Pole: A Sequel to 'From the Earth to the Moon'* (London, 1891).

7 Denis Cosgrove, *Apollo's Eye: A Cartographic Genealogy of the Earth in the Western Imagination* (Baltimore, MD, 2001), pp. 49, 109.

8 Ibid., pp. 34–5; Virgil, *The Sixth Book of the Aeneid*, ed. A. J. Church (London, 1872).

9 Godwin, *Arktos*, pp. 106–7.

10 Margaret Cavendish, *Observations upon Experimental Philosophy: To which is added, The Description of a New World, Called The Blazing-world* (London, 1666).

11 James Fitzmaurice, 'Cavendish, Margaret, Duchess of Newcastle upon Tyne', *Oxford Dictionary of National Biography* (Oxford, 2004); Katie Whitaker, *Mad Madge: Margaret Cavendish, Duchess of Newcastle, Royalist, Writer and Romantic* (New York, 2002); John Rogers, *The Matter of Revolution: Science, Poetry, and Politics in the Age of Milton* (Ithaca, NY, 1996); Emma L. Rees, *Margaret Cavendish: Gender, Genre, Exile* (Manchester, 2003).

12 Cavendish, *The Blazing World*.

13 Ibid., p. 7.

14 Stephen Clucas, ed., *A Princely Brave Woman: Essays on Margaret Cavendish, Duchess of Newcastle* (Aldershot, 2002).

15 Mary Shelley, *Frankenstein, or, the Modern Prometheus* (London, 1818); Laurie Garrison, 'Imperial Vision in the Arctic: Fleeting Looks and Pleasurable Distractions in Barker's Panorama and Shelley's *Frankenstein*', *Romanticism and Victorianism on the Net*, 52 (November 2008), available at www.erudit.org, accessed 1 October 2017; Tim Fulford, Debbie Lee and Peter Kitson, *Literature, Science, and Exploration in the Romantic Era: Bodies of Knowledge* (Cambridge, 2004), pp. 149–76.

16 Samuel Taylor Coleridge, 'The Rime of the Ancient Mariner', in *Lyrical Ballads* (London, 1798); Anon., *Munchausen at the Pole* (London, 1819); Poe, *Arthur Gordon Pym*.

17 For Regency and Victorian audiences, narratives of Arctic exploration acquired a powerful and complex array of meanings. Adriana Craciun, *Writing Arctic Disaster: Authorship and Exploration* (Cambridge, 2016); Robert David, *The Arctic in the British Imagination, 1818–1914* (Manchester, 2000); Russell A. Potter, *Arctic Spectacles: The Frozen North in Visual Culture, 1818–1875* (Seattle, WA, 2007); Janice Cavell, *Tracing the Connected Narrative: Arctic Exploration in British Print Culture, 1818–1860* (Toronto, 2008).

18 Boyd Hilton, *Corn, Cash, Commerce: The Economic Policies of the Tory Governments, 1815–1830* (Oxford, 1977); Boyd Hilton, *The Age of Atonement: The Influence of Evangelicalism on Social and Economic Thought, 1785–1865* (Oxford, 1986); James Grande and John Stevenson, eds, *William Cobbett, Romanticism and the Enlightenment* (London, 2015).

19 Christopher Lloyd, *Mr Barrow of the Admiralty: A Life of Sir John Barrow, 1764–1848* (London, 1970); Craciun, *Writing Arctic Disaster*; Adriana Craciun, 'Writing the Disaster: Franklin and Frankenstein', *Nineteenth-Century Literature*, LXV/4 (2011), pp. 433–80.

20 Helmut Müller-Sievers, *The Science of Literature: Essays on an Incalculable Difference* (Berlin, 2015).

21 Thomas C. Dundonald, *The Autobiography of a Seaman* (London, 1861); Christopher Lloyd, *Lord Cochrane: Seaman – Radical – Liberator* (London, 1947).

22 Brian Vale, *Cochrane in the Pacific: Fortune and Freedom in Spanish America* (London, 2008).

23 Thomas McLean and George Cruikshank, 'The Sailor's Progress: "Arrival at the North Pole"', etching from the series *The Progress of a Midshipman exemplified in the Career of Master Blockhead* (London, 1835).

24 *Munchausen at the Pole*, p. 5.

25 Ibid., p. 21.

26 Ibid., p. 93.

27  Ibid., p. 94.

28  Ibid.

29  John Gascoigne, *Science in the Service of Empire* (Cambridge, 1998); Sophie Waring, 'The Board of Longitude and the Funding of Scientific Work', *Journal for Maritime Research*, XVI/1 (2014), pp. 55–71.

30  Shane McCorristine, *The Spectral Arctic: A Cultural History of Ghosts and Dreams in Polar Exploration* (London, 2018).

31  Ibid.

32  Heidi Hansson, '*Punch, Fun, Judy* and the Polar Hero: Comedy, Gender and the British Arctic Expedition, 1875–76', *North and South: Essays on Gender, Race and Region*, ed. Christine DeVine and Mary Ann Wilson (Newcastle, 2012), pp. 61–90.

33  Mark Duell, 'The North Pole Bid that Never Was: Forgotten Photos from Failed 1875 British Expedition Reveal the Scurvy-hit Crew's Efforts 30 Years Before History was Made', *Mail Online*, 20 September 2016, www.mailonline.com, accessed 14 December 2017.

34  Thomas Ingham, 'Exploring a Heritage: Tracing Victorian Thoughts on Ancient Exploration', BA diss., Department of Classics, University of Cambridge, 2017.

35  Fridtjof Nansen, *In Northern Mists: Arctic Exploration in Early Times* (London, 1911).

36  Adolf E. Nordenskiöld, *Facsimile Atlas to the Early History of Cartography*, trans. Johan A. Ekelöf and Clements R. Markham (London, 1889).

37  Robert E. Peary, 'The Lure of the North Pole', *Pall Mall Magazine*, XXXVIII (1906), pp. 342–5.

38  Felix Driver, *Geography Militant: Cultures of Exploration and Empire* (Oxford, 2000); Karen Morin, *Civic Discipline: Geography in America, 1860–1890* (Farnham, 2011).

39  William F. Warren, *Paradise Found: The Cradle of the Human Race at the North Pole: A Study of the Prehistoric World* (Boston, MA, 1885), p. 71; Brook Wilensky-Lanford, *Paradise Lust: Searching for the Garden of Eden* (New York, 2011).

40  Godwin, *Arktos*, p. 8.

41  Warren, *Paradise Found*, p. 52.

42  Bal G. Tilak, *The Arctic Home in the Vedas: Being also a New Key to the Interpretation of many Vedic Texts and Legends* (Poona, 1903).

43  A.A.M. [A. A. Milne], 'An Unconvincing Narrative', *Punch*, CXXXVII (15 September 1909), p. 188.

44  My gratitude to Heidi Hansson for her guidance with the contested readings of Milne's polar writing. See Heidi Hansson, 'Nordpolen enligt Puh: Alternativa arktiska diskurser i brittiska

populära framställningar 1890–1930 [The North Pole According to Pooh]', *Reiser og Ekspedisjoner i det Litterære Arktis*, ed. Johan Schimanski, Cathrine Theodorsen and Henning Howlid Wærp (Trondheim, 2011), pp. 239–61.

45  A. A. Milne, *Winnie-the-Pooh* (London, 1926), p. 112.

46  Ibid., p. 122.

47  Ann Thwaite, *A. A. Milne: His Life* (London, 1990).

## 6  Sovereigns of the Pole

1  A central text in this debate that spawned many valuable responses is Keith Thomas, *Religion and the Decline of Magic: Studies in Popular Beliefs in Sixteenth- and Seventeenth-century England* (London, 1971).

2  Michael Robinson, 'Reconsidering the Theory of the Open Polar Sea', in *Extremes: Oceanography's Adventure at the Poles*, ed. Keith Benson and Helen Rozwadowski (Sagamore Beach, MA, 2006), pp. 15–30.

3  'Umberto Nobile', https://en.wikipedia.org, accessed 1 October 2017.

4  Adriana Craciun, *Writing Arctic Disaster: Authorship and Exploration* (Cambridge, 2016).

5  Wilbur Cross, *Disaster at the Pole: The Tragedy of the Airship 'Italia' and the 1928 Nobile Expedition to the North Pole* (Guilford, CT, 2000).

6  One of many North Pole expeditions illustrating the fine line between hubris and determination was the expedition of Walter Wellman; see P. J. Capelotti, *The Greatest Show in the Arctic: The American Exploration of Franz Josef Land, 1898–1905* (Norman, OK, 2016).

7  Jules Verne, *The Purchase of the North Pole: A Sequel to 'From the Earth to the Moon'* (London, 1891)

8  Verne, *Purchase*, pp. 7–8.

9  Ibid., p. 13.

10  Lee A. Farrow, *Seward's Folly: A New Look at the Alaska Purchase* (Fairbanks, AK, 2016).

11  Verne, *Purchase*, p. 34.

12  Ibid.

13  Ibid., p. 47.

14  Ibid., pp. 47–8.

15  James Macadam, 'The Arctic Coal Rush: Spitsbergen and the British Imagination, 1910–1920', MPhil diss., University of Cambridge, 2011; Richard Drayton, *Nature's Government: Science, Imperial Britain, and the 'Improvement' of the World* (New Haven, CT, 2000).

16 Verne, *Purchase*, pp. 50–51.

17 Joseph Moxon, *A Brief Discourse of a Passage by the North-Pole to Japan, China, etc. Pleaded by Three Experiments . . .* (London, 1674), pp. 1–2.

18 John Dixon Hunt, 'Emblem and Expressionism in the Eighteenth-century Landscape Garden', *Eighteenth-century Studies*, IV/3 (1971), pp. 294–317.

19 Beau Riffenburgh, *The Myth of the Explorer: The Press, Sensationalism, and Geographical Discovery* (London, 1993).

20 Robert E. Peary, 'The Lure of the North Pole', *Pall Mall Magazine*, XXXVIII (1906), pp. 342–5.

21 Ibid., p. 344.

22 *Plato's Sophist: Part II of 'The Being of the Beautiful'*, trans. S. Bernadete (Chicago, IL, 1984), p. II.71.

23 Peary, 'Lure of the North Pole', p. 344.

24 Eric G. Wilson, *A Spiritual History of Ice: Romanticism, Science, and the Imagination* (New York, 2003).

25 Peary, 'Lure of the North Pole', p. 343.

26 Ibid.

27 Robert E. Peary, *The North Pole: Its Discovery in 1909 under the Auspices of the Peary Arctic Club* (New York, 1910), p. 260.

28 Peary, 'Lure of the North Pole', p. 342.

29 Ibid., p. 344.

30 Peary, *The North Pole: Its Discovery*, p. 259.

31 Ibid.

32 Peary, 'Lure of the North Pole', p. 342.

## 7 Mourning Antaeus

1 Marionne Cronin, 'Technological Heroes: Images of the Arctic in the Age of Polar Aviation', in *Northscapes: History, Technology, and the Making of Northern Environments*, ed. Dolly Jørgensen and Sverker Sörlin (Vancouver, 2013), p. 58.

2 Robert N. Rudmose-Brown, 'Recent Polar Work – Some Criticisms', *Polar Record*, I/5 (1933), pp. 62–6. After participating as the botanist on William Speirs Bruce's Scottish National Antarctic Expedition (1902–4), he took up an academic position at Sheffield University. By the time of writing his 1933 article, he was recognized as one of Britain's authorities on the polar regions.

3 Ibid., p. 62.

4 Ibid., p. 64.

5 Ibid.

6 This perspective draws on the classic essay of Walter Benjamin, 'The Work of Art in the Age of Mechanical Reproduction', trans. J. A. Underwood (London, 2008)

7   Vilhjalmur Stefansson, *The Friendly Arctic* (London, 1921), p. 8.
8   Christina Adcock, 'Tracing Warm Lines: Northern Canadian Exploration, Knowledge and Memory, 1905–1965', PhD diss., University of Cambridge, 2010.
9   Stefansson, *Friendly Arctic*, pp. 1–26.
10  Bronislaw Malinowski, *Argonauts of the Western Pacific: An Account of Native Enterprise and Adventure in the Archipelagoes of Melanesian New Guinea* (London, 1922).
11  *Nanook of the North: A Story of Life and Love in the Actual Arctic* (dir. Robert J. Flaherty, 78 minutes, 1922).
12  Stefansson, *Friendly Arctic*.
13  Ibid., p. 8.
14  Vilhjalmur Stefansson, 'The Region of Maximum Inaccessibility in the Arctic', *Geographical Review*, x/9 (1920), p. 168.
15  Aleksandr Kolchak, 'The Arctic Pack and the Polynya', *Problems of Polar Research*, ed. W.L.G. Joerg (New York, 1928) pp. 125–41. Kolchak, a Russian admiral, identified the geometric centre of the ellipse-shaped polar ice pack in 1909, probably the first formulation of a Pole of Inaccessibility, but he didn't share Stefansson's later concern with the biogeography of living on the sea ice.
16  Stefansson, *Friendly Arctic*, p. 19.
17  David Turnbull, 'Maps, Narratives and Trails: Performativity, Hodology and Distributed Knowledge in Complex Adaptive Systems – an Approach to Emergent Mapping', *Geographical Research*, xlv/2 (2007), pp. 140–49.
18  Takashi Irimoto and Takako Yamada, *Circumpolar Religion and Ecology: An Anthropology of the North* (Tokyo, 1984).
19  Hugh Brody, *Living Arctic: Hunters of the Canadian North* (London, 1987); Barry Lopez, *Arctic Dreams: Imagination and Desire in a Northern Landscape* (New York, 1986).
20  Julie Cruikshank, *Do Glaciers Listen? Local Knowledge, Colonial Encounters and Social Imagination* (Vancouver, 2005).
21  Stefansson, 'Maximum Inaccessibility', p. 172.
22  Lincoln Ellsworth, *Beyond Horizons* (New York, 1938), p. 217.
23  Sverker Sörlin, ed., *Science, Geopolitics and Culture in the Polar Region: Norden Beyond Borders* (Farnham, 2013).
24  Antonio Quattrini, 'Titina: The Odyssey of a Terrier Waif', *New York Times*, 9 January 1927, pp. 35, 161.
25  William Anderson, *Nautilus 90 North* (London, 1959), p. 156.
26  Colin Bertram, *Arctic and Antarctic: The Technique of Polar Travel* (Cambridge, 1939).
27  Jean de Pomereu, 'The Exploration of *Inlandsis*: A Cultural and Scientific History of Ice Sheets to 1970', PhD diss., University of Exeter, 2015; Victoria Herrmann, 'Layered Landscapes: Space,

Place, and Power in Alaska, America's Last Frontier (1942–2015)',
PhD diss., University of Cambridge, 2018.

28  Elmer Plischke, 'Trans-polar Aviation and Jurisdiction over Arctic
Airspace', *American Political Science Review*, xxxvii/6 (1943), p. 1000.

29  *Life*, 19 July 1943, pp. 8–9.

30  Plischke, 'Trans-Polar Aviation', p. 999.

31  Arctic Council, *Impacts of a Warming Arctic* (Cambridge, 2004).

32  See for example Pen Hadow's 90*North Unit's campaign to protect
the wildlife and ecoystem of the international waters surrounding
the North Pole, www.penhadow.com, accessed 15 May 2018.

33  June Rose, *Marie Stopes and the Sexual Revolution* (London, 1992).

34  Morgan Seag, 'Women Need Not Apply: Gendered Institutional
Change in Antarctica and Outer Space', *Polar Journal*, vii/2 (2017),
pp. 319–35; Morgan Seag, 'Equal Opportunities on Ice: Examining
Gender and Institutional Change at the British Antarctic Survey,
1975–1996', MPhil diss., University of Cambridge, 2015.

35  Embla Eir Oddsdóttir, Atli Már Sigurðsson and Sólrún Svandal,
*Gender Equality in the Arctic: Current Realities, Future Challenges*
(Reykjavik, 2015).

36  'About Ann Bancroft', www.annbancroftfoundation.org, accessed
15 October 2017.

37  Jason Daley, 'The Amazing Story of the First All-women North
Pole Expedition', 11 July 2017, www.smithsonianmag.com, accessed
16 December 2017.

38  'Barbara Hillary (explorer)', https://en.wikipedia.org, accessed
16 December 2017.

39  Dolores Johnson, *Onward: A Photobiography of African-American
Polar Explorer Matthew Henson* (Washington, DC, 2005).

40  *Sea Blind: The Price of Shipping Our Stuff* (dir. Sarah Robertson and
Bernice Notenboom, 61 minutes, 2015).

41  Hilde Elise Heidal and Hilde Kristin Skjerdal, 'Radioactive
Contamination in the Barents Sea', Institute of Marine Research,
2 September 2015, www.imr.no, accessed 16 December 2017.

42  Andrés Cózar et al., 'The Arctic Ocean as a Dead End for Floating
Plastics in the North Atlantic Branch of the Thermohaline
Circulation', *Science Advances*, iii/4 (2017), e31600582.

43  Daniel Bravo, personal communication on the significance of the
golden apples of the Hesperides, October 2017.

44  Seamus Heaney, 'Antaeus', in *North* (London, 1975), p. 3.

# SELECT BIBLIOGRAPHY

Bennett, Jim, *The Divided Circle: A History of Instruments for Astronomy, Navigation, and Surveying* (Oxford, 1987)

Berggren, J. Lennart, and Alexander Jones, *Ptolemy's Geography: An Annotated Translation of the Theoretical Chapters* (Princeton, NJ, 2000)

Boyle, Deborah, *The Well-ordered Universe: The Philosophy of Margaret Cavendish* (New York, 2017)

Brotton, Jerry, *A History of the World in Twelve Maps* (London, 2012)

Capelotti, Peter, *By Airship to the North Pole: An Archaeology of Human Exploration* (London, 1999)

Cosgrove, Denis, *Apollo's Eye: A Cartographic Genealogy of the Earth in the Western Imagination* (London, 2001)

Craciun, Adriana, *Writing Arctic Disaster: Authorship and Exploration* (Cambridge, 2016)

Crane, Nicholas, *Mercator: The Man who Mapped the Planet* (London, 2002)

Davidson, Peter, *The Idea of North* (London, 2016)

Drivenes, Einar-Arne, and Harald Dag Jølle, *Into the Ice: The History of Norway and the Polar Regions* (Oslo, 2006)

Godwin, Joscelyn, *Arktos: The Polar Myth in Science, Symbolism, and Nazi Survival* (London, 1993)

Heaney, Seamus, *North* (London, 1975)

Imbert, Bertrand, *North Pole, South Pole: Journeys to the Ends of the Earth* (New York, 1992)

Ingold, Tim, *The Perception of the Environment: Essays on Livelihood, Dwelling and Skill* (Abingdon, 2000)

Jones, Kathleen, *A Glorious Fame: The Life of Margaret Cavendish, Duchess of Newcastle, 1623–1673* (London, 1988)

Leane, Elizabeth, *South Pole* (London, 2016)

Leask, Nigel, *Curiosity and the Aesthetics of Travel Writing, 1770–1840: 'From an Antique Land'* (Oxford, 2002)

McCannon, John, *A History of the Arctic: Nature, Exploration and Exploitation* (London, 2012)

McCorristine, Shane, *The Spectral Arctic: A Cultural History of Ghosts and Dreams in Polar Exploration* (London, 2018)

MacDonald, John, *The Arctic Sky: Inuit Astronomy, Star Lore, and Legend* (Toronto, 1998)

Malaurie, Jean, and Sylvie Devers, *Pôle Nord 1983: Histoire de sa conquête et problèmes contemporains de navigation maritime et aérienne* (Paris, 1987)

Morin, Karen, *Civic Discipline: Geography in America, 1860–1890* (Farnham, 2011)

Potter, Russell A., *Arctic Spectacles: The Frozen North in Visual Culture, 1818–1875* (Seattle, 2007)

Pumfrey, Stephen, *Latitude and the Magnetic Earth* (Cambridge, 2002)

Robinson, Michael, *The Coldest Crucible: Arctic Exploration and American Culture* (Chicago, IL, 2006)

Romm, James, *The Edges of the Earth in Ancient Thought: Geography, Exploration, and Fiction* (Princeton, NJ, 1992)

Scafi, Alessandro, *Mapping Paradise: A History of Heaven on Earth* (London, 2006)

Snyder, John, *Flattening the Earth: Two Thousand Years of Map Projections* (Chicago, IL, 1993)

Sumira, Sylvia, *Globes: 400 Years of Exploration, Navigation, and Power* (London, 2014)

Wilson Rowe, Elana, *Arctic Governance: Power in Cross-border Cooperation* (Manchester, 2018)

Woodward, David, *The History of Cartography*, vol. III: *Cartography in the European Renaissance*, 2 vols (Chicago, IL, 2007)

# ACKNOWLEDGEMENTS

I would like to thank the many people without whose invaluable support this book would not have been possible. In the first instance, Michael Leaman, my publisher, and Daniel Allen, the Earth Series editor, wrote to me suggesting the topic. The deft guidance of my editor Amy Salter has been invaluable, as has the excellent work of my picture editor, Susannah Jayes. As a historian of science working at a polar institute, I was struck by how little I knew about the North Pole, what species live there, what resources are found there or whether it is simply a mathematical fiction, a sort of non-place. I even wondered whether there was enough to say to justify a book on the pole in this series. The volume of material discovered in the course of research, with the help of colleagues, unearthed an enormous amount of seldom-discussed material. This has forced me to be highly selective and I apologize to readers disappointed to find many expeditions overlooked, Santa Claus absent and that the North Pole today is mentioned only in passing.

The inspiration for taking the long view of the North Pole came from my friendship with Denis Cosgrove. A very special thanks to Megan Barford for assistance with picture research, and Elana Wilson Rowe for extensive editorial advice. For critical discussion of draft chapters, I thank Adriana Craciun, John MacDonald, Simon Schaffer, Morgan Seag, Sylvia Sumera and David Turnbull, as well as colleagues at symposia at the Clark Library, the Cabinet of Natural History in the Cambridge Department of the History and Philosophy of Science, the National Maritime Museum, and the Social Sciences Department at the Arctic University of Norway.

Historical geographers Stephen Daniels, Felix Driver, Luciana Martins, Catherine Nash, Miles Ogborn and Charles Withers have been lasting influences, as have Bill Adams, Ash Amin, Matthew Gandy, Emma Mawdsley, Clive Oppenheimer, Susan Owens and Liz Watson in the Cambridge Geography Department. At the Scott Polar Research Institute, my social science colleagues and students, former and present, have also been my teachers: Christina Adcock, Henry Anderson-Elliott, Mia Bennett, Marionne Cronin, Peter Evans, Janne Flora, Penny Goodman, Johanna Grabow, Huw Lewis-Jones, Bryan Lintott, Shane McCorristine, Richard Powell, Jackie Price, Gareth Rees, Morgan Seag, Piers Vitebsky, Claire Warrior and Corine Wood-Donnelly. Thanks also to the Institute's director, Julian Dowdeswell, as well as to Martin French and Peter Lund at the Institute's Library, Naomi Boneham at the Thomas H Manning Polar Archives, and Charlotte Connelly, Rosie Ames and Naomi Chapman at the Polar Museum. My thanks also to the Place, Space and Mobilities Research Group at the Arctic University of Tromsø, particularly Anniken Førde, Berit Kristoffersen, Britt Kramwig, and Torill Nyseth, as well as Harald Dag Jølle and Mary Jones (formerly) at NorskPolarinstitut for generously sharing their historical knowledge.

I'm also indebted to Jim Bennett, Barbara Bodenhorn, Robert Marc Friedman, Heidi Hansson, Jonathan Lamb, Ruth Maclennan, Katie Parker, Anka Byall, Stephen Pumfrey, Nicky Reeves, Rolf Schneider, Richard Staley, Sverker Sörlin, Ruth Stirling and Urban Wråkberg.

I would like to acknowledge the generous support for this research from the University of Cambridge's Humanities and Social Sciences Research Grant Scheme, as well as a grant from the Royal Society's History of Science Scheme. I am also grateful to the Wissenschaftskolleg, Berlin, for their kindness and encouragement in offering me the opportunity to spend a fellowship period of ten months with my family to develop the ideas in this book.

Finally, thanks go to my wife, Emma, and children, Mia, Daniel and Esme, for their love and support and for believing passionately in books.

# PHOTO ACKNOWLEDGEMENTS

The author and the publishers wish to express their thanks to the below sources of illustrative material and /or permission to reproduce it.

Alamy: pp.31 (Heritage Image Partnership Ltd), 38, 47 (World History Archive), 40–41, 44 (prisma archivo), 62–3, 132–3, 137, 179 (Granger Historical Picture Archive), 66 (Northwind Picture Archive), 82 (The Print Collector), 104 (ART Collection), 142 (Chronicle), 143 (World History Archive), 144 (Science History Images), 161 (Hemis), 164 (Pictorial Press), 189 (Photo Researchers), 205 (marka), 209 (sputnik); Author's Collection: pp. 184, 190, 187, 198; With permission of the Bodleian Library, University of Oxford: p. 33 (MS.Marsh.144); Michael Bravo: pp. 21, 23, 26, 28; British Library, London: pp. 154, 155, 156; By permission of the British Museum, London: pp. 150–51; Reproduced by kind permission of the Syndics of Cambridge University Library: pp. 6 (688.c.91.4) , 57 (Atlas.3.55.2), 58 (Atlas.3.55.2 ), 64 (Adams.6.55.1), 67 (Adams.6.55.1), 69 (Adams.6.55.1), 105 (Map.bb.17.F.1) 108 (Gg.34.21), 109 (Gg.34.21), 115 (Mm.34.15), 116 (Q.382.c.66.2), 117 (E.33.19), 118 (Ll.24.5), 120 (Ll.24.5), 122 (Ll.24.5), 126 (Q340.1.c.33.5), 127 (340:01.a.1.6), 160 (Atlas 3.55.2), 162 (Atlas 3.55.2), 165 (8460.d.220), 167 (P*.5.10(E)), 172 (1890.7.2046), 174 (1890.7.2046) ,177(1875.7.523), 188 (L996.c.38), 192 (L900.b.94), 201 (P690.b.37), 210 (RCS.Per.1081); Miles Cudmore: p. 215; By permission of Ann Daniels: pp. 218, 219; Courtesy of Dartmouth College Library: p. 197; Geomatics and Cartographic Research Centre, Carleton University (GCRC): p. 22; Getty Images: pp. 50, 163 (DeAgostini Picture Library), 83, 149 (Science & Society Picture Library), 152 (Photo by Archiv Gerstenberg/ullstein build via Getty Images), 158 (The Cartoon Collector/Print Collector), 178 (Bettmann), 183 (Mansell/ The LIFE Picture Collection); Martin Hartley: pp. 220, 221; Mike King: pp. 213, 214, 216; Reproduced by kind permission of the Middle Temple Library: pp. 11, 76; National Gallery of Art, Washington DC: pp. 180, 181;

# INDEX

Page numbers in *italics* refer to illustrations